D1509317

The Genetic Perspective

Adela S. Baer
San Diego State University

W. B. SAUNDERS COMPANY
Philadelphia, London, Toronto

W. B. Saunders Company: West Washington Square
Philadelphia, PA   19105

1 St. Anne's Road
Eastbourne, East Sussex BN21 3UN, England

1 Goldthorne Avenue
Toronto, Ontario M8Z 5T9, Canada

Cover illustration, showing a portrait of Gregor Mendel, his glasses and a drawing of sweet peas, courtesy of the Corning Glass Works, Corning, New York.

The Genetic Perspective   ISBN   0-7216-1471-X

© 1977 by W. B. Saunders Company.   Copyright under the International Copyright Union.   All rights reserved.   This book is protected by copyright.   No part of it may be reproduced, stored in a retrieval system, or transmitted in any form or by any means, electronic, mechanical, photocopying, recording, or otherwise, without written permission from the publishers.   Made in the United States of America. Press of W. B. Saunders Company.   Library of Congress Catalog card number 76-014672.

Last digit is the print number: 9   8   7   6   5   4   3

Dedication
For Susan and Nicoli Baer

PREFACE

Taken as a whole, this book allows the inquisitive mind to delve into most aspects of how heredity affects us and the rest of the living world. Through it, one can become acquainted with the science of genetics and become conversant with the way geneticists think and talk about their special branch of science. The book may also help the reader to relate his or her personal observations on the nature of life to the grand biological schemes generated by flurries of study and analysis.

In addition, the book provides glimpses of other fields of science and the humanities as seen from the geneticists' window on the universe. These glimpses show the mutual dependence that exists between genetics and other sciences, and also between genetics and nonscientific human concerns. However, the book does not provide an encyclopedic view of genetics or of biology generally. To obtain that view requires an enduring association with more detailed genetic-biology texts.

As this book attempts to show, possessing some understanding of genetics can illuminate the biological relationships within the human realm and also the role of human beings in the natural world — the oneness of humanity and the oneness of life. Moreover, having some understanding of genetics can increase one's aesthetic awareness of nature's diverse patterns. This diversity enriches our world. For example, it furnishes us with the joys and entertainment of human individuality, which is a reflection of the unpredictable genetic uniqueness of every fertilized egg.

I came to write this book after much discussion about genetics in its scientific and social context with students, friends, and colleagues in several countries. Such discussion has been a welcome "occupational hazard" for me. Although many of the people I talked to are not scientists, they have shown an eloquent curiosity about the human significance of heredity, as well as curiosity about the ardent absorption of geneticists with the subject of genetics. I have benefited greatly from their persistent questions and fresh points of view. This is not surprising; genetics has become a core area in biology, in large part by absorbing any relevant information that came into view. Perhaps because of these omnivorous activities, some geneticists hold that an understanding of genetic principles provides the passkey to the rest of biology. At any rate, geneticists have ventured boldly into surrounding territory, such as statistics, entomology, biochemistry, ornithology, argiculture, medicine, and anthropology. (Of course, the traffic has been two-way; other disciplines frequently come to genetics to gather ingredients for their theories.)

In summary, genetics is usually in ferment, absorbing ideas and methods from other sciences and from occasional "passers-by," such as students and philosophers. Genetics has, I think, enough flexibility, openness, and charm to assure it of a healthy future (barring nuclear war) into that distant age when our children's children's children will no longer remember us by name, only by our lively genes.

ACKNOWLEDGMENTS

It has been my pleasure to have had the critical aid of Carol Barnett, George Cox, Jack Cronin, Mary Fitzgerald, Duane Jeffery, Skai Krisans, Richard Lampert, Candy McCombs, Roberta Palmour, Jean Passino, Rollin Richmond, and June Shepard in the preparation of this book.

CONTENTS

The Genetic
Perspective

NOBEL WINNER RAPS STUDY OF DISEASES

MIAMI, Fla. (AP) — Dr. Francis Crick, the British biologist who shared a 1962 Nobel Prize for the discovery of the DNA molecule, said America spends too much time and money on medical research.

"Americans have a peculiar illusion that life is a disease that has to be cured," Crick said in an interview following a lecture at the University of Miami.

"Everyone gets unpleasant diseases," added Crick, whose work with DNA, the basic chemical building block of the human body, has been hailed as the most significant biological research of the 20th century. "And everyone dies at one time. I guess they are trying to make life safe for senility."

In America, Crick said, scientists relate to what kind of research brings the biggest money.

"One year it's cancer," he said. "The next year it's old age. Before we can cure, we have to understand how nature work

Crick shared the Nobel Prize with American molecular biologist Dr. J. D. Watson.

New Artificial Gene Reported by MIT Team

Scientists from the Massachusetts Institute of Technology said Wednesday they have successfully developed an artificial gene which could, with a little work, function inside

Disease Potential

Genetic Peril in 'Miracle' Crops

By Charles Petit

The fact that science can create new strains of plants in a test tube—at the same time the world's natural gene pool is being depleted—offered a strange scenario yesterday for the possible future of evolution.

Two reports, delivered during the 13th International Congress of Genetics in Berkeley, illustrated the widely-varied paths being taken by plant geneticists, especially breeders of food

netic base to resist disease, he said. But in the long run, Frankel added, the greater loss is the genetic information that could have been saved for future plant breeding.

EXTINCTION

The same situation holds for wild plants and animals, he said, as man-made changes drive local varieties, even entire species, to extinction. Without proper safeguards such as adequately-maintained national

You are what you eat.

...gy and stamina, ...ant thing you must ...ght. Good-tasting

cell plants back into complete organisms, leaves and all.

He has, for instance, come

All Men Are Created Equal---but How About Women?

Sterile Male Rats Make for 'Zero Growth'

The rats in Jones, Okla., are being wiped out---not by guns or pesticides, but by birth control. More than a year ago specially bred sterile male rats were introduced into the rat population of the Oklahoma City suburb, giving females false pregnancies ...ake them nonproductive for months. And Jones' rats are diminishing, says Dr. Allan Stanley, a research scientist, who started the federally supported program.

The Jones experiment was begun with a rat-population count in a rural area at the edge of town. The area then was divided into two sectors with equal rat populations. Laboratory-bred sterile rats were substituted for wild rats in one sector. Then, rat-population counts were made in both sectors, and the sector with laboratory rats showed a decline. The genetically flawed sterile rats cut the sector's rat population to 30% of its former level in a year

...y snare my ...ecause she ...than I en...also perhaps ...premonition ...would come

...ue, in the case ...ly loses often ...ra...bb and strokes ...m...age and tutor...as in a Sunday ...r level usually ...t that way with ...ed to watch the ...d pull for Riggs ...nething for me. I

...s so lopsided it was ...sing. Riggs dinked ...placed and topspun, ...something, though

tion. Although dealing directly with military service and the relative standing of men and women "under arms," the court felt no obligation to publish any internal argument over whether women are just as good infantrypersons as men.

Wherever the Equal Rights Amendment is at issue, its opponents have warned that in removing all legal discrimination against women, it would make them eligible for the military draft when there is one. And carried further, it would make them liable for combat assignments without regard to their sex.

The theory is that since most women cannot lift as much or run as fast, and have less violent dispositions, they are unsuited to combat duty. Never mind my own memory that the biggest men in any outfit

In ruling as it did on dependents' support, the court ignored that, because it is beside the point.

Riggs beat Court, and probably will beat Billie Jean King, too. If I am lucky I will keep beating my favorite opponent. Chances are Mark Spitz will keep his swimming records until some other man, not woman, breaks them. But all that is beside the point, too.

The point is that it is not necessary to *be* equal to be treated equally. This came up last year when a Nobel laureate claimed scientific proof that blacks are intellectually inferior. I see contrary evidence all around me, but even if he were correct, that also would be immaterial.

If anything is unanimously accepted in our country, it should be

Another sort of people banishes chance altogether and attributes events to a star or to their horoscopes, believing for all men, born and unborn, God's decree has once and for all been enacted so that for the rest of time he may sit and rest and sit still. This belief begins to take root among both the educated and the ignorant mob who march rapidly in that direction. Witness the warnings drawn from lightning, the forecasts made by oracles, the predictions of soothsayers and ridiculous trifles—a sneeze or a stumble counted as omens.... Thus these things deceive and entangle mortals so that all that remains certain is that nothing certain exists.

Pliny the Elder, A.D. 23 to A.D. 79

Chapter One

BELIEFS ABOUT HEREDITY

Myths, superstitions, and prejudices associated with biological inheritance and human development are common in most cultures. Prejudices about heredity often are involved with mistrust of strangers or of outside social groups. Superstitions and myths often are inspired by the surprising appearance of an abnormality in a member of the family. Superstitions about such abnormalities may arise from a coincidence—eating a deformed carrot sometimes precedes having a deformed child. We now know some of these abnormalities are hereditary. They are caused by genes, the chemical blueprints in the body that determine structures and functions.

Biological inheritance is the realm of study of geneticists. Geneticists study how traits are inherited, what substances are involved in their transmission and expression, and how inherited traits relate to medical problems and other aspects of our society.

In scientific-industrial societies, people are generally well informed about sex and biological inheritance. At least, they know how offspring are produced by mating. Nevertheless, there are still some misconceptions about why offspring look and behave in the unpredictable ways they do. The teaching of genetic principles at many levels of education is gradually overcoming much of this confusion. Certain aspects of this confusion, which I call genetic fads and fallacies, are discussed in this chapter.

COMMON QUESTIONS ABOUT HEREDITY

Are All Our Social Problems Environmental?

The main problem with such a question is the common fallacy of "single cause" thinking. Heredity and environment are interdependent; they are not separate entities. Since heredity, or genes, are never independent of *some* social and physical environment, all our social problems, human miseries, and human joys are dependent on both our heredity and our environment. We are all visibly and genetically different (except for identical twins), and, in addition, no two people have the same experiences, food preferences, or medical ailments. Therefore, it is clear that not everyone can become a musician, a football player, or a mother. But the interactions between our heredity and our environment are, unfortunately, unknown in most cases. Social problems such as health, education, and mental stress depend on our biological abilities as well as our cultural attitudes, that is, on both our nature and our nurture (see Chaps. Fourteen and Nineteen).

Is There a Genetic Basis for Criminality?

This question, once popular, went out of style for several decades because geneticists did not find any firm evidence to support this notion. Recently, it has popped up again as a result of the discovery that a few men have some excess genetic material: they have an extra copy of what is called the Y chromosome. Apparently, excessive numbers of these men were found in prisons, which has led to speculation that the extra chromosome predisposed them to aggressive acts that society considers to be criminal. The details of this story are given in Chapters Six and Fourteen. Suffice it to say here that men with an extra Y chromosome have been found who have never been in prison or been labeled aggressive by society. This genetic condition has therefore been discredited as a cause of criminality. Men with this genetic condition vary greatly in their behavior, perhaps as much as everyone else.

Are Genetic Defects Incurable?

Some genetic defects can be treated. Near-sightedness, for example, can obviously be corrected by wearing glasses; these treat the symptoms but do not cure the ailment. Whereas other defects, such as colorblindness, are not yet treatable, the proportion of inherited abnormalities that one "can't do anything about" is decreasing rapidly. Physicians now routinely treat with some success certain inheritable defects that were impossible to alleviate or cure in the recent past. Among these are diabetes, phenylketonuria, and hemophilia.

Alleviating or curing a person's symptoms, however, is not enough. The problem hidden from view here is that the "cure" doesn't cure the afflicted person's genes. When you reproduce, your defective genes can be passed on to some of your children, and the problems of misery and treatment are compounded in the next and the next generations. Where all this leads is discussed in Chapter Twenty.

Do Drugs Cause Genetic Abnormalities?

The drug awareness of the United States population has inspired a flurry of research reports on the effects of popular drugs. Few commonly abused drugs have been shown to produce detectable genetic effects. While some studies have concluded that LSD, lysergic acid diethylamide, does not cause genetic or other defects, other studies have concluded that it does. Since the publication of these alarming reports, LSD usage has declined nationally, perhaps because of them. In a recent laboratory study, marihuana was reported to cause damage to chromosomes, the physical sites of the genes. But a study of long-term heavy consumers of marihuana does not uphold this finding. Other drugs, such as alcohol, as well as food preservatives and pollutants, receive less publicity but are viewed more suspiciously by geneticists (see Chap. Thirteen).

Are Birth Defects Due to Shock or Stress on the Mother?

Congenital defects are ones present from birth onward. Some have unknown causes. Others are known to be caused by certain drugs during pregnancy, by infection (for instance, German measles or syphilis), or by harmful inherited factors, that is, by genes. The mother's state of physical health and her nutrition during pregnancy are very important to the well-being of an infant. Poor nutrition alone can produce a retarded, anemic, or bone-deformed child. Fetal development may also be influenced by exposure of the pregnant mother to radiation (see Chap. Thirteen).

No one has found strong evidence for attributing birth defects to maternal emotional distress. Yet, as we learn more about the biological basis of emotions and behavior, we might find that stressful, or even joyful, emotions of a mother have *some* effect on the growth and disposition of her unborn child. Abundant evidence, on the other hand, does confirm that acquired changes, such as a mother's toes being cut off, do not influence the offspring. If a newborn infant is missing a toe, geneticists postulate other explanations, for example, the effects of deformity-causing drugs such as thalidomide, or the effects of mutation or recessive genes (see Chap. Two).

Are Human Beings Becoming Less Intelligent?

No clear decline in intelligence over the generations has been demonstrated. In fact, human intelligence may still be increasing. The perennial problem in this regard is that researchers cannot agree on how to define or measure intelligence. The size of the brain case has clearly increased, from the fossil remains of our ancestors to the human species today. This certainly indicates a dramatic increase in human cleverness in the past. Although human beings are not the largest of the primates, we have the largest brains. We may indeed be the brainiest creatures the earth has ever known—but to assert this statement *strongly* might be taking a narrow-minded view of the life styles of dolphins, elephants, and other beasts that are known to have large brains and to be clever in their own ways.

Is It Possible To Produce a Genetic Superman or Super Race?

Perhaps by some selective breeding program or by genetic surgery (Chap. Twenty) we could grow super people. But is that really desirable? Who decides what is "super"? Uncertainties in artificial insemination technology, social problems in rewarding or coercing reproduction by selective "breeders," and other hazardous matters are involved. While Chapter Twenty explores this topic more fully, let us here consider a counterquestion: With adequate nutrition, freedom, and medical care, most people everywhere could have a zestful, energetic, responsible, thoughtful, and productive existence: what more could a society ask for? Genetic and other utopias are imagined by people who do not fully appreciate the genetic and other real complexities of man and nature.

Are Women Genetically Inferior to Men?

Women do have more gene material than men, a factor which might contribute to their greater potential for a longer life. On the other hand, women are on the average smaller and physically weaker than men. These facts do not signify that women are either genetically superior or inferior to men overall; the two sexes are just genetically different (see Chaps. Five and Six).

When people discuss genetic differences between the sexes, they sometimes pounce on the question of differences in intelligence between men and women. There may indeed be differences in both the quality and kinds of intelligence men and women possess, but the evidence is not compelling one way or the other (see Chap. Fourteen). The basic problem in this controversy is to devise convincing tests to identify and compare intelligence of all sorts; this problem has not yet been solved.

What is "Good Breeding"?

People who consider themselves "aristocrats" may wish to believe they are biologically superior in order to justify continuing their privileged way of life. The idea of genetic inferiority by class, race, or religion is a political one and, in its popular meaning, is contradicted by biological facts. This point is discussed extensively in later chapters.

Is Inbreeding Bad?

Matings between close relatives are not necessarily bad. One must consider the context of "badness" and ask the question, *harmful for survival under what conditions?* Racehorses and isolated human groups on Pacific islands are quite inbred, but they are both fully capable of survival in their own particular environments. Inbred populations are, however, more likely than large populations to have a peppering of inherited defects (see Chap. Three).

The other side of the question on inbreeding is sometimes emphasized when it is stated that hybrids, which are produced by outbreeding, are categorically "good." While hybrid corn is very productive, and thus very popular, present-day intensive farming of hybrid crops is not, for various reasons, "good" overall (see Chap. Eighteen). For human beings, however, there is no evidence that hybrids are either more or less vigorous than anyone else. The old habit of calling hybrids "half-breeds" is misguided because some people think the term implies "half value," and misleading because it implies that there are pure-bred groups. Nothing could be further from the truth. Human populations have been extensively hybridized through the ages. Invaders invade, explorer groups explore, pilgrims progress—all this adventure inevitably adds new genes to local populations.

What Is a Blood Relative?

Though it is a figment of folklore, the term "blood relative" is still found in everyday speech. Most people presumably do not mean it literally. Blood is not an ingredient in the production of offspring. An egg cell from the mother and a sperm cell from the father combine to form a fertilized egg which grows and transforms into the embryo, fetus, and baby. The genes in the fertilized egg determine the form and function of the various tissues and organs in the new individual. It is valid, therefore, to speak of "gene relatives" or "genetic relatives."

Good blood, or bad blood, does not exist in hereditary terms. Aristocrats do not have some special blue blood that would be diluted by letting a commoner into the family. In the nineteenth century a man named Galton transfused blood between black and white rabbits to confirm the popular belief of inheritance through blood. The rabbits mated, white with white and black with black. Galton expected the offspring to be spotted or perhaps gray. Nothing of the sort happened. The white parents had white offspring and the black ones had black.

Blood transfusions have no effect on one's heredity or the heredity of one's offspring. One modern example of "blood will tell" is the rare situation of blood-type incompatibility between mother and fetus in which fetal blood accidentally leaks into the mother's circulation. This genetic incompatibility does not alter the child's heredity, but may hurt its chance of survival (see Chap. Eleven).

SUGGESTED READINGS

Freire-Maia, A., C. Stevenson, and N. E. Morton. Hybridization effect on mortality. Social Biology 21:232–234, 1974.

Hays, H. R. Birds, Beasts, and Man. Baltimore: Penguin Books, 1972.

Hook, E. B., and W. J. Schull. Why is the XX fitter? Nature 224:43–46, 1973.

Medawar, P. B. The Future of Man. New York: New American Library, 1961.

Robin, E. D. The evolutionary advantage of being stupid. Perspectives in Biology and Medicine 16:369–380, 1973.

Stern, C. Principles of Human Genetics. Ed. 3. San Francisco: W. H. Freeman & Co., 1973, p. 429.

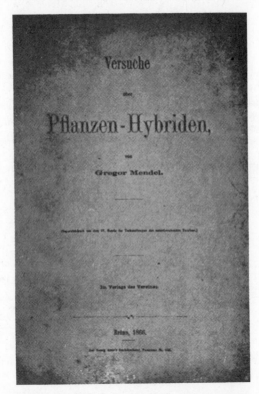

From Horowitz, N. H.: The Gene. Scientific American, October, 1956.

Chapter Two

HOW TRAITS ARE INHERITED

Much of science has arisen through efforts to solve practical problems. For genetics, the practical reality that guided early workers was in agriculture; their concerns embraced the breeding of plants and animals and the effects of hybridization between breeds, the choice of crops adapted to certain climates, and the like. People who had practical experience in such matters were the ones who often raised the most perceptive questions. Gregor Mendel was one of these (see Box on Mendel, p. 10). He and Charles Darwin, whom we shall discuss later, were two of the towering geniuses of nineteenth-century biology.

MENDEL'S STUDIES

By experimentally crossing strains of pea plants, Mendel showed in 1865 how variable characteristics are inherited. His findings established a major biological concept: an inherited trait is caused by a specific unit of life substance which is transmitted intact from parent to offspring. Mendel called this unit of substance a factor; now it is called a gene. To understand how the gene concept came into being, we shall consider Mendel's ideas, his experimental design and results, and his own logical analysis of his findings.

Mendel chose to study garden peas because many varieties with clear-cut differences, such as the color of the flower or seed, were available to him from farmers. (One biologist I know has suggested that Mendel worked on garden peas simply because he liked to eat them, but this is sheer conjecture.) The other advantages of working with pea plants were as follows: they reproduce every year; they are cheap and easy to grow, requiring little gardening care; they are not very susceptible to insect damage; and they produce many seeds (peas in the pea pod). Pea flowers contain both male and female sex cells (called pollen and ovules), and pea offspring are normally produced by self-fertilization (see Fig. 2–1). That is, before the flower opens and lets in pollen from another plant, the ovules in the base of the flower are fertilized by its own pollen. Cross-fertilization, on the other hand, can be used to produce a hybrid artificially: to this end Mendel brushed the pollen from one plant inside the developing flower of another plant before its own pollen was ripe. (In many other plants, however, hand-pollination is difficult or impossible because of the smallness or the complexity of the flowers.)

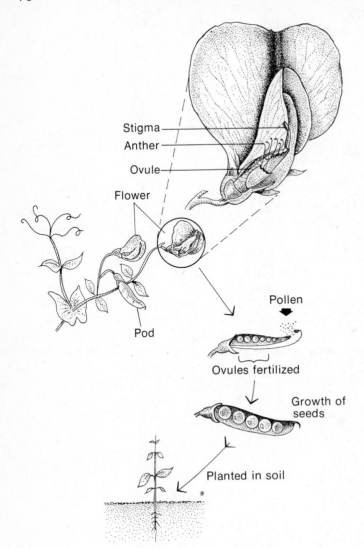

Stigma
Anther
Ovule
Flower
Pod
Pollen
Ovules fertilized
Growth of seeds
Planted in soil

Figure 2-1 Mendel's peas. Close-up diagram shows the structures that make artificial pollination possible.

BIOGRAPHY OF MENDEL (1822–1884)

Gregor Mendel was born in 1822 about 75 miles north of Vienna. Vienna was then the capital of a large empire which included present-day Austria and Hungary, as well as parts of Germany, Italy, Yugoslavia, Rumania, and Czechoslovakia (where Mendel lived). This multi-ethnic empire, repressively ruled by an aristocracy that survived until the First World War, was feudalistic (while other European nations and the United States already had constitutions). Feudalism was outlawed in the empire, during Mendel's lifetime, only after riots in Vienna and unrest elsewhere. Because the Hapsburg royal family was Catholic, the clergy led a privileged existence; this was to benefit Mendel by providing him, as a monk, with leisure for pursuing scientific studies.

The son of a peasant, young Gregor learned the art of farming that was later to prove useful in his scientific studies on plant hybrids. His family, including his sisters, made willing sacrifices to send him to school. In high school he was taught "natural history" (biology) and some physical science; this curriculum, extremely modern for the time, was imposed on the school with the determination of Countess Waldberg, a local aristocrat.

All his life Mendel had a studious bent. He decided to become a priest in order to pursue his scholarly interests without financial hardship to his family. At the age of 21 he joined the Augustinian monastery in the nearest town (Brno, Czechoslovakia). Five years later he was ordained a priest, and the following year he became a high school substitute teacher, a job he held off and on for many years. Although he wanted to be a regular teacher, he failed the licensing examination—a curious defeat that may have resulted from a combination of too much reliance on self-teaching and too independent a frame of mind regarding biological theories.

When Mendel was 29 his abbot sent him, at the monastery's expense, to the University of Vienna, where he pursued his fascination with science. Here for two years Mendel studied physics, zoology, plant physiology, and mathematics. Upon returing to the monastery he devoted his efforts to botanical studies in the cloister gardens. For seven years (1856–1863) he toiled in the summer sun studying the hybrids of garden peas. It is this work which showed that heredity follows a logical pattern and which laid the foundation for the science of genetics. He reported on this work to his local natural history society in 1865 and published his findings in the society's journal the next year.

No one seems to have found Mendel's studies of interest then or during the remainder of his lifetime, despite the fact that he wrote to a famous botanist explaining them. His results were, however, "discovered" in the year 1900 by later plant geneticists. One possible reason for the earlier neglect of Mendel is this: while he was looking at how individual traits are inherited, other biologists were looking at how the whole complex of traits that characterize a species changes selectively over time. This widespread interest in the evolution of species was precipitated in 1859 by the publication of Charles Darwin's book *On the Origin of Species* (see Chap. Seventeen).

In 1868, when Mendel was 46 years old, he was elected abbot of his monastery. Thereafter, although he kept abreast of scientific developments through the literature, he had little time for his own scientific studies. He became a conscientious administrator and a stout defender of policies of non-taxation of monasteries (at a time when the government was experiencing acute financial distress). He waged this battle against taxes for 12 years. He fell ill and died in 1884.

Experimental Design.   Instead of trying to keep track of all the characteristics of pea plants at once, Mendel decided to study one characteristic at a time. He picked out characteristics — such as the color of the peas in the pod — that were distinctively different among his chosen varieties. Thus he could distinguish varieties at a glance, rather than have to contend with particular plants that were "more or less" like one variety or another. Moreover, he studied only those traits that were not changeable, that is, traits whose expression was not noticeably influenced by the amount of rain, sunshine, soil conditions, or other environmental fluctuations.

The characteristics Mendel chose for study were seven in number. These included the surface texture of the seed (smooth or wrinkled), the height of the plant (tall or dwarfed), and the color (yellow or green) of the pea-seed "meat" (which botanists call cotyledons). Armed with these carefully chosen, stable characteristics of peas, Mendel was then able to "ask" his experiments how such a simple difference as yellow versus green seeds was inherited by offspring from their parents.

What Mendel did was, in retrospect, quite simple and obvious. (The same sorts of experiments are done every year by students in biology classes.) First, he grew distinctively different strains of garden peas in separate plots for two years to observe if they would "breed true." That is, he let them self-fertilize two seasons to see if they remained constant in their traits. Those strains that faithfully reproduced their traits both years were selected for further study.

Let us consider two of these selected strains that differ in the color of their seeds, one having yellow and the other green. (Dried yellow and green peas are still available in many grocery stores.) By artificial cross-pollination Mendel produced hybrid seeds in plants of the green-seeded variety using pollen from a yellow-seeded variety — thus performing the cross, green female × yellow male. Because of his propensity for meticulousness he also made hybrids by the reciprocal cross (yellow female × green male). Figure 2–2 charts the results of these crosses in a flow diagram. The initial cross between parental green and parental yellow is designated P × P.

Results.   In both crosses, Mendel found, only yellow seeds were produced in the hybrid generation — called the first filial or F_1 generation. He concluded that the yellow effect simply dominated over the green alternative in the F_1 since it made no difference if the yellow trait came from the pollen (male) parent or the ovule (female) parent. Genticists say that Mendel showed that "yellow is dominant to green" in peas.

Mendel stored these F_1 yellow seeds during the winter separately from the original, true-breeding yellow seeds. Because no one could have distinguished the two kinds of yellow seeds by looking at them, we say they have the same phenotype. But Mendel knew that they had different histories: the F_1 yellow seeds had one green-seeded parent, but the original, true-breeding yellows had only yellow-seeded parents (and other ancestors). So he now set out to determine if the F_1 yellows would behave differently from the true-breeding yellows in the genetic future — possibly because of this known difference in their genetic past.

As Mendel knew, other investigators had also observed the dominant inheritance of a parental trait, such as yellow, in the F_1 hybrid offspring of a cross between different strains. But few had pursued the matter further. Mendel studied the genetic future of his F_1 yellows by growing plants from the F_1 seeds the following year, letting them self-fertilize and waiting for harvest time to observe what seed colors they produced. Would they be like the original yellow-seeded variety and upon "selfing" breed true (produce only yellow seeds) or not?

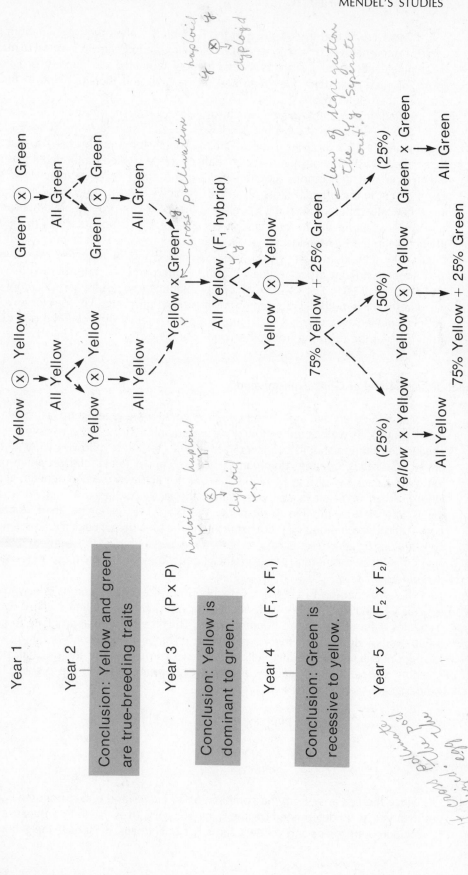

Figure 2-2 A flow diagram charting the results of Mendel's experiments (⊗, self-fertilization).

Mendel found that the offspring of the F_1 hybrids, called the F_2 generation, included representatives of both phenotypes of their grandparents. In contrast to many other nineteenth-century plant breeders, Mendel tabulated the number of F_2 offspring of each type (both the green and the yellow) and then resolved to figure out what the results meant.

He had grown 258 F_1 plants which, by self-fertilization, produced 8023 seeds (F_2's): 6022 seeds were yellow and 2001 were green. This is a ratio of about 3 to 1 (approximately 75% to 25%). The first striking conclusion from this result is that the "green gene" was obviously not lost or changed in the F_1 seeds, although it was not manifest in that generation. Apparently the green effect survived intact from the green parent in the original cross (P $\times$ P). This shows that the P generation yellows are genetically different from the F_1 yellows: the latter can produce green offspring upon selfing, while the former cannot. Because of the reappearance of the green phenotype in the F_2, Mendel called the green gene recessive with respect to yellow. In F_1 yellows, the dominant yellow gene is expressed and the recessive green is not.

Another striking aspect of the F_2 results is the particular ratio observed. Why 3 to 1 instead of 10 to 1 or 1 to 1? After contemplating these results, Mendel proposed a model of, or hypothesis for, gene transmission to offspring that, as it turned out, not only made sense of the 3 to 1 ratio, but also correctly predicted the ratios one would get for the offspring from other kinds of crosses.

Mendel's Analysis and Model of Gene Transmission

Mendel argued that since the F_1 yellows contain the green gene (as shown by the F_2 results), as well as the yellow gene (producing their yellow phenotype), their genetic constitution should be symbolized as follows: Y (for yellow) + y (for green), or Yy. (Geneticists denote dominant genes by capital letters and recessive ones by lower case.) Recall that Mendel's F_1's did not act as if the two genetic factors were blended or fused in any way; rather, the factors behaved as individual elements in their transmission to offspring. We say the F_1 genotype, then, contains both Y and y, even though y is not expressed in the phenotype. Such a mixed genetic constitution, or genotype, is said to be heterozygous. That is, the zygote (fertilized egg) contains two different [hetero (other)] genetic elements or, more technically, two different alleles.

The idea that the F_1 contains a double dose of alleles is compatible with the general fact that two parents are needed for reproduction: pollen without ovules or ovules without pollen are nonproductive. Thus it would be appropriate to have a two-symbol designation for the genotypes of other individuals. We can use YY for the original, pure-breeding yellows; and we can use yy for the original, pure-breeding greens. Both YY and yy are homozygous (i.e., alike in the zygote). Thus we designate Mendel's first hybrid cross:

parental phenotypes: yellow $\times$ green
genotypes: YY   yy
F_1 phenotype: yellow
genotype: Yy

Since the F_2's arise from the combination of ovules and pollen from the F_1, the problem is now to determine how the F_1 genotype Yy gives rise to F_2's. Because the F_2's include both greens and yellows, some F_2's (the greens) must have the genotype yy, and some (the yellows) must have a genotype containing at least one Y. But are the F_2 yellows YY or Yy? Both genotypes, you recall, *look* yellow.

Segregation of Alleles. Mendel answered this question by his model of segregation of alleles to the reproductive cells. He suggested that the reproductive cells, or gametes, are single-factored in genotype: the gametes produced by the Yy F_1 plant can have Y or y, but not both. By postulating that any single F_1 reproductive cell had an equal chance (50:50) of containing Y or y, Mendel was able to show that both YY and Yy genotypes were expected in the F_2 and also that 75% yellow to 25% green was the correct ratio.

Under his scheme, when F_1 heterozygotes are selfed their pollen will be 50% Y and 50% y, and their ovules will likewise be 50% Y and 50% y. If, as Mendel assumed, male reproductive cells fertilize ovules only through chance encounters—an occurrence known as "random union of gametes"—then we will have:

Yy × Yy   } F_1 genotypes

F_1 pollen

	Y	y
Y	YY	Yy
y	yY	yy

F_1 ovules   } F_2 genotypes

The diagram above is called a Punnett square, a device invented by R. C. Punnett about 60 years ago to visualize offspring proportions in Mendelian crosses. In a Punnett square the genotypes of the parental gametes are written on the top and on one side, and the allelic combinations of offspring are symbolized inside the square (see Fig. 2–3).

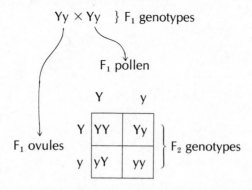

Figure 2–3 Simple Mendelian crosses.

Mendel's F_2 genotypic ratio of 1 to 2 to 1 (1:2:1) perfectly explains his ob-served 3 to 1 (3:1) phenotypic ratio of yellow to green because both the 1/4 YY and 1/2 Yy are yellow. (We sometimes write this ratio as 3Y–:1yy, where the dash in-dicates the second position, which can be occupied by either allele.) That is,

F_1 yellows: Yy × Yy
F_2 genotypes: 1/4 YY + 2/4 Yy + 1/4 yy
phenotypes: 3/4 yellow + 1/4 green

Mendel tested this hypothesis about the different genotypes among the F_2's by letting them reproduce by self-fertilization in a subsequent year. He found what he ex-pected: (1) all F_2 greens gave only green offspring in the F_3 (yy × yy → all yy); (2) 1/4 of the yellow F_2's gave only yellow F_3 offspring (YY × YY → all YY); and (3) 2/4 of the yellow F_2's gave a ratio of 3 yellow to 1 green in their offspring (Yy × Yy → 3/4 Y– + 1/4 yy).

PROBABILITY

Mendel drew upon his knowledge of simple mathematics to explain his experiments. In the cross Yy × yy, the offspring were 50% Yy and 50% yy, as discussed in the text. This 1:1 ratio implies that for any given offspring the chance is 50% it will be Yy and 50% it will be yy. The odds, if we were going to bet on the outcome, for pulling either kind of offspring ''out of a hat'' are 1 to 1. There is an even chance that we will pull out Yy or yy.

Similarly, with a 3:1 ratio of yellow to green from the cross Yy × Yy, the odds, or chances, are 3 to 1 that any one offspring we pick will turn out to be yellow. If someone offered even money on this, the bettor that picked yellow every time would win 75% of the time and the bettor that stayed with green would win only 25% of the time.

Figuring the odds in genetics (as in playing poker or betting on horses) is based on a certain logic. In genetics the logic says that, given the segregation of alleles to different gametes in heterozygotes (50% containing Y and 50% containing y), the chance of getting a yy individual from the Yy × Yy cross is 50% × 50%, or 1/2 × 1/2. This is true because the two events (y-bearing ovule and y-bearing pollen) are independent. We say that the probability, or chance, of two independent events occurring simultaneously is the product (not sum) of their independent probabilities. Thus for the probability of obtaining yy from Yy × Yy we figure 1/2 × 1/2 = 1/4 (not 1/2 + 1/2 = 1). What is the probability of YY from this same cross? What is the probability when you flip two coins simultaneously that both will come up heads?

In algebra, the general expression for these cases is called the binomial expansion. Applied to the Yy × Yy Mendelian cross, this expression reads $(1/2\ Y + 1/2\ y)^2 = 1/4\ YY \times 1/2\ Yy + 1/4\ yy$.

By this F_3 test (one of several possible progeny tests), Mendel verified his model of allele segregation to gametes and the subsequent random union of gametes to form zygotes. Since one of Mendel's hallmarks was thoroughness, he tested his model of allele transmission to offspring on the case of a cross-fertilization between the F_1 yellow (Yy) and a green (yy). He predicted the offspring of this cross would be 1/2 yellow and 1/2 green as a result of segregation of the Y and y alleles to separate gametes in the Yy parent and the subsequent random union of gametes. His results were just as expected (see Box on Probability, p. 16).

parent phenotypes:	F_1 yellow × green
genotypes:	Yy yy
progeny phenotypes:	1/2 yellow + 1/2 green
genotypes:	Yy yy

That is,

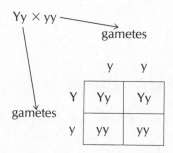

In all the cases Mendel studied, reciprocal crosses gave the same results, as stated earlier. We shall see, however, that sometimes it does matter whether it is the mother or father that carries a certain gene; this is true for X-linked genes, which are discussed in Chapter Five.

Further discussion of the diverse patterns of gene transmission is given in Appendix A.

THE IMPORTANCE OF MENDEL'S WORK

Mendel came to understand basic principles of heredity by making sense out of simple observations. His experiments were well thought out beforehand and patiently carried through to completion. Investigators of heredity before Mendel's studies had fallen short of a clear understanding because they failed to be systematic and thorough in their experimentation. As Mendel pointed out in his research report, they did *not* (1) keep the similar phenotypes of the different generations separate; (2) note the number of phenotypic classes in each generation; (3) analyze the numerical relations of the phenotypic classes found; or, (4) follow out the crosses for enough generations. Mendel methodically did all of these. He thus came to appreciate that heredity, is determined by discrete elements for particular traits, and that these elements are regrouped each generation in a predictable fashion. Mendel's model of the segregation of alleles to gametes and their random union in zygotes was, and is, a simple and elegant explanation for biological inheritance. His conclusions were a triumph of reasoning. Predictions that flow from his ideas have been verified for many different organisms. (See Box on Scientific Problem Solving, p. 18.)

SKEPTICISM ABOUT MENDEL'S ACCURACY

In every generation of students who read about Mendel, there are inevitably a few who point out that Mendel's results agree almost too well with his expected 3:1 and 1:1 ratios. Considering the rarity of getting *exactly* 50 heads and 50 tails in 100 flips of a coin, it is hard to disagree with these skeptics. If Mendel did fudge his results to fit his theories—for which no evidence exists among Mendel's preserved papers—he may have done so unconsciously (and therefore undeliberately) simply by stopping his tally of, say, yellow vs. green seeds when he felt he had "counted enough" and had obtained a "good" representative sample. Such unconscious workings of the human mind are a well-known danger to objectivity in science and other endeavors.

Although many have questioned the extreme accuracy of Mendel's results, such skepticism does not extend to the validity of the results. Mendel's evidence for, and concepts of, gene transmission remain the unchallenged foundation of genetics.

SCIENTIFIC PROBLEM SOLVING

Mendel's work is widely regarded as exemplifying clear thinking about a problem and clear writing about it. Few scientists are so successful in seeing a problem and solving it satisfactorily. Many scientific discoveries probably occur by methods that are poorly understood even by the scientists involved. Luck, hunches, and dogged perseverance all play their part at times. But the methods Mendel used fall within the realm of the widely accepted "scientific method." In general, the methods of science are methods congenial to all human minds. Many people use scientific reasoning and problem-solving techniques in everyday life. Scientists, however, use these techniques perhaps more routinely than non-scientists do. The education of scientists emphasizes the study of the methods and reasoning used by other workers in one's field of interest.

A scientist often has an idea of the type, "I wonder if phenomenon X is caused by Z." He may then read the published results of other scientists on the subject and, if necessary, reformulate his guess, or hypothesis, about the cause. His hypothesis at this point should be compatible with all the previous information on the subject and it should predict the outcome of observations or experiments that so far exist only as a gleam in the scientist's eye. The testing of the hypothesis may involve either observations of natural occurrences (experiments in nature) or manipulations of natural occurrences (experiments in the laboratory, the greenhouse, or the fishtank).

Mendel hypothesized that the inheritable characteristics of pea plants could be studied separately—that one could study a single characteristic apart from all the others in a particular plant and that this single trait would have a simple, discrete determinant somewhere in the makeup of the living material of the plant. Mendel's hypothesis predicted that the simple determinant (the gene) would have different effects in different states or combinations. If the trait of seed color has two possible states (Y or y) for its gene, and if each plant has two such genes, then there are three possible combinations: YY, Yy, and yy. Mendel must have reasoned that, if this hypothesis were true (1) there would be a 3:1 phenotypic ratio in the F_2 from the $F_1 \times F_1$ cross when Y is dominant to y; (2) some of the F_2 yellows would be heterozygous and segregate out greens in the F_3; and so on. This form of argument—called deduction—is perhaps most famous as the kind of reasoning purported to be used by Sherlock Holmes.

Mendel's deductions from his hypothesis turned out to be valid: the predictions were borne out by his results. No evidence has since been found by later scientific workers to contradict Mendel's conclusions on the particulateness of genes and their segregation to gametes. But we do know of interesting exceptions to Mendel's findings that have helped us to understand patterns of gene transmission in great detail.

STUDY QUESTIONS

1. What phenotypes would you expect in the pea progeny of the cross: F_1 yellow $\times$ pure-breeding yellow? What genotypes would you expect?

2. Another trait Mendel studied in pea plants was height. When he crossed a pure-breeding tall variety to a pure-breeding dwarf, all the F_1's were tall. The F_2's obtained by selfing the F_1's were about 75% tall and 25% dwarf. What is the dominant gene here? What is the expected proportion of homozygous tall in the F_2? Of homozygous dwarf? What is the genotypic ratio in the F_2? Which phenotype is the recessive one?

3. The Englishman T. A. Knight used artificial cross-pollination to improve the characteristics of grapes, plums, cherries, and garden plants experimentally. He reported on crosses in peas in 1823 (when Mendel was just one year old). Knight found that a yellow-seeded variety crossed to a green-seeded variety gave all yellow-seeded offspring. When the F_1 yellows were allowed to self-fertilize, they produced both yellow and green F_2's. Knight did not count the proportions of yellow and green F_2's as Mendel did. Why is it important to determine these proportions (the phenotypic ratio)?

4. Breeding better crops is one important application of genetic principles (see Chap. Eighteen). Recently the nutrient value of corn has been improved by the introduction of a "high lysine" gene into varieties of field corn. The gene gives corn a better balance of protein for the human diet. If a true-breeding "high" strain is crossed to a true-breeding "low" strain, what phenotype will the F_1 have if the allele for high is recessive (i.e., H = low, h = high)? What must be the parents' genotypes and phenotypes for a cross that produces a high-to-low phenotype ratio of 1:1 among the progeny?

5. Leeuwenhoek, the famous microscope lover who first saw living sperm cells and other microscopic life, wrote to the Royal Society of London in 1683:

> . . . Many of our neighbors either for their pleasure or profit keep tame rabbits which are long eard, ordinarily of a white color but sometimes of blew, black, and pyed. Those that would make a profit of these rabbits by causing them to bring grey young ones, which in the spring may be sold for a wild kind, get a grey male, such as are ordinarily found on our sand hills, to put to their female. The breed that comes from hence always takes the grey color of the father, and it has never been seen that any of the young has any white, or other colored hair than grey, there withal they are never so big as the dam, nor have so great ears, nor are so tame, but of a wilder kind.

If we assume "the breed that comes from hence" refers to the F_1, what colors would be expected in the F_2 if a neighbor of Leeuwenhoek lost his white female (and had eaten his gray male) and decided to keep and mate together the gray F_1's produced from a cross of white times gray?

6. St. Isidore, writing about A.D. 630, declared:

> . . . in herds of horses, those of noble birth are brought [by men] before the sight of those conceiving, by means of which they are able to conceive and create offspring resembling the noble stock. . . . Hence it is that certain people command their pregnant women not to look upon the shameful countenances of animals such as . . . monkeys lest they would bear offspring like those they met in their vision. This is the nature of females that whatever they look at they take up without thought in the extreme ardor of passion and bring forth offspring of that kind.

Discuss this passage.

7. Words and phrases important for understanding Mendelian genetics:

gene	F_1	heterozygote
true-breeding	F_2	homozygote
hybrid	dominant	gamete
self-fertilization	recessive	allele
cross-fertilization	phenotype	allele segregation
progeny	genotype	
reciprocal cross	zygote	

8. If human beings could self-fertilize as garden peas do, how would you characterize the genotypes of future generations—as becoming increasingly heterozygous or homozygous? Why?

SUGGESTED READINGS

Dunn, L. C. A Short History of Genetics. New York: McGraw-Hill Book Co., 1965.

Gasking, E. B. Why was Mendel's work ignored? Journal of the History of Ideas 20:62–84, 1959.

Harpstead, D. D. High-lysine corn. Scientific American 225:34–42, 1971.

Jenkins, J. Genetics. Boston: Houghton Mifflin Co., 1975.

Olby, R. C. Origin of Mendelism. New York: Schocken Books, Inc., 1966.

Orel, V. The scientific milieu in Brno during the era of Mendel's research. Journal of Heredity 64:314–318, 1973.

Stern, C. Principles of Human Genetics. Ed. 3. San Francisco: W. H. Freeman & Co., 1973. Chapter 6 contains a good discussion of probability.

Stern, C., and E. R. Sherwood (eds.). The Origin of Genetics: A Mendel Source Book. San Francisco: W. H. Freeman & Co., 1966.

Stubbe, H. History of Genetics. Cambridge, Mass.: MIT Press, 1972.

Sturtevant, A. H. A History of Genetics. New York: Harper & Row Publishers, Inc., 1965.

Zirkle, C. The Beginnings of Plant Hybridization. Philadelphia: University of Pennsylvania Press, 1935.

Velázquez: Las Meninas. The woman behind the dog is an achondroplastic dwarf. This is a dominant trait. From Museo del Prado, Madrid. Used by permission of the Museo del Prado, Madrid.

Chapter Three

HUMAN HEREDITY

GENES IN FAMILIES

It has long been observed that particular traits run in families. Many family traits are caused by genes which are passed on from parents to children in egg and sperm cells. One curious but harmless trait that is occasionally seen in various societies is a white forelock (Fig. 3–1). In families in which a white forelock occurs it is usually passed on from generation to generation in an unbroken line—grandfather George having it, one daughter and perhaps one son having it, and then, say, the daughter's daughter having it. Geneticists, taking their cue from genealogical studies of Scottish clans and Mayflower descendants, diagram such white forelock families as follows:

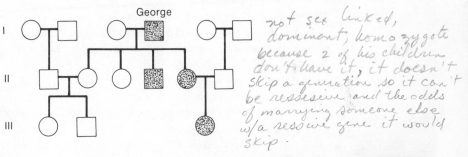

(handwritten note:) not sex linked, dominant, homozygote because 2 of his children don't have it, it doesn't skip a generation so it can't be ressesive and the odds of marrying someone else w/a ressive gene it would skip.

Figure 3–1 Woman with a white forelock.

23

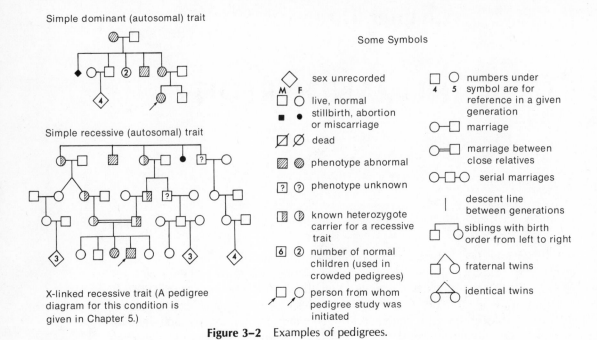

Figure 3–2   Examples of pedigrees.

By traditional agreement the circles stand for females and the squares for males in such pedigree diagrams. Darkened symbols represent persons showing the trait in question. Roman numerals designate the generations. Generalized pedigree diagrams are given in Figure 3–2.

Note that in the forelock pedigree on page 23, and indeed usually when a trait runs in a family, not all the offspring of an affected parent show the trait. For example, since George is heterozygous (genotype Aa) for the forelock and his wife is homozygous normal (genotype aa), half their children are expected to have white forelocks (a 1:1 ratio). Here, in fact, two of the four siblings in the family do have the forelock trait. With characteristics like white forelock (and other dominant traits), family members and geneticists alike are impressed with the fact that when *neither* parent shows the trait, their children do not show it either: all of them (genotype aa) lack the dominant allele for the forelock, as shown for two children in generation III. The genotypic diagram for George's pedigree looks like this:

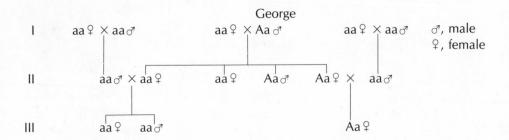

Unlike white forelock, however, many inherited traits are not harmless. The list of harmful genetic traits in humans has already exceeded 1000. This list includes acromegaly (distortions of facial and other bones), albinism, some eye cataracts, some forms of deafness, hemophilia (bleeder's disease), polydactyly (extra fingers or toes), abnormalities of sexual development, sickle-cell anemia, Tay-Sachs disease (a form of infant idiocy), xeroderma pigmentosum (a kind of skin disease), and many others. Fortunately, most of these traits are quite rare. Moreover, some of them show up every generation and some skip generations; that is, some are dominant and some recessive.

Albinism.   Because of their lack of skin pigmentation, albinos are very sensitive to sunlight; they are prone to sunburn and skin cancer. (See Fig. 3–3.) A lack of pigment in the eye makes them also prone to blindness. Albinos crop up sporadically in many cultural groups. (See Box on the Winter brothers, p. 27.) Figure 3–4 shows a pedigree diagram of albinism in the Hopi Indians of Arizona. Notice in the diagram that the albino boy in generation VI has no albino relative closer than his grandfather's brother. This skipping of generations suggests that albinism is a recessive trait; at least that is the simplest possible genetic explanation. (Recall that the green-seeded trait of Mendel's peas skipped the F_1 generation.) The effect of albino alleles becomes manifest only when normal alleles for pigmentation are absent in the individual—that is, when the albino alleles are homozygous.

The albino phenotype is much more common in a few population isolates than it is in large human populations. For example, in 1967 there were 28 albinos living among the approximately 6000 Hopi Indians on their reservation in Arizona. This frequency—approximately one albino among every 200 Hopi—stands in sharp contrast to that of one albino among 20,000 in white populations. The high frequency of Hopi albinos is thought by geneticists to be due in part to inbreeding: albino children are much more common in cousin marriages than in marriages in

Figure 3–3   Hopi women and girls, Mishongnovi, 1898. There are two albinos in the group. (From R. Mahoud, ed.: Photographer of the Southwest: A. C. Vroman. Pasadena, Ward Ritchie Press, 1969. Used by Permission.)

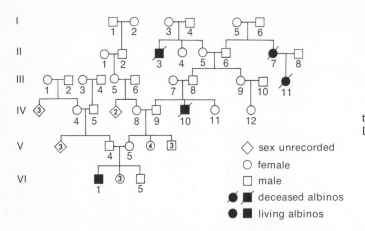

Figure 3–4 Pedigree of albinism in the Hopi Indians (courtesy of F. C. Dukepoo).

◇ sex unrecorded
○ female
□ male
●⬟ deceased albinos
●■ living albinos

which the parents are not related. This is so in all cultures because two cousins often are heterozygous for the same recessive allele, which both inherited from some common ancestor. The prohibition on marriages between close relatives in various societies makes sense in this regard: inbreeding increases the chance of recessive defects showing up.

CELLS, CHROMOSOMES, AND GENE CHEMISTRY

The human body is composed of packets of life substances called cells. All cells, including skin cells, liver cells, nerve cells, and, of course, fertilized egg cells, contain genes. Genes consist of chemical material packaged in bodies called chromosomes. In each cell the chromosomes are sequestered within a special compartment known as the nucleus. This much has been known since about 1900 (see Chap. Four); but the chemical makeup of genes has been understood only since the 1950's.

The gene substance, deoxyribonucleic acid, is known as DNA. Geneticists now understand how a trait like albinism can be determined by a minute piece of chemical material—DNA—inside the body cells (see Chap. Seven). Within the cells, many genes determine the production of individual proteins, some of which act as enzymes, the biological catalysts for biochemical reactions. Albinos lack the enzyme necessary for converting food materials into dark pigment in the skin and hair. This enzyme deficiency in albinos occurs because they did not receive any gene from their parents that contained the correct chemical (DNA) instructions for making the enzyme protein. Without a correct genetic blueprint, their cells cannot construct the enzyme and thus cannot manufacture skin pigment. Nor can albinos pass on to their offspring correct DNA instructions for synthesizing this pigment-producing enzyme, since none of their cells contain such DNA. Other individuals, however, who have the proper DNA instructions for making the pigment-producing enzyme— that is, who have a normal gene—can both manufacture pigment and pass on the correct genetic instructions to their offspring.

POPULATIONS OF GENES

Just why there are both albinos and normally pigmented people in populations is a tricky question. We know that albinos occur in many animals (and plants): mice, deer, tigers, and birds, for example, as well as humans. Laboratory studies on plants and animals show clearly that once in a great while an albino mouse or

JOHNNY AND EDGAR WINTER AND INDIVIDUALITY

Johnny and Edgar Winter are brothers who are successful muscians who combine blues, jazz, and rock styles. Both are albinos. When the brothers were growing up in Beaumont, Texas, their parents, both of whom were talented in music, encouraged their musical interests. Johnny, the elder of the two sons, was extroverted, whereas Edgar was shy. "I didn't have many friends," Edgar says. "You know the way kids naturally are if you're fat, crippled, or in any way defective. They tend to leave you out. So music became my identity and replaced the normal activities that otherwise would have filled my life."

Edgar Winter. (Photo courtesy of Columbia Records.)

Like many albinos, Edgar is almost totally blind because of abnormal development of eye tissues. This problem kept Edgar from participating in sports but contributed to his learning how to play music by ear in virtuoso fashion.

As adults, the brothers have received solid recognition in the world of entertainment, in part because of their albinism. "I used to perform, not communicate," Edgar says. "Now I accept audiences and that has overcome my withdrawal. I'm much happier now and better balanced. Being albino always gave me a very real sense of individuality. Today, in music, a lot of people will do anything to themselves just to set them apart. I guess I've had a natural edge on them."*

*Quoted material from J. F. Jerome: For a song. People, Vol. 2, No. 4 (July 22, 1974), 53–54.

corn plant "crops up" in a group that had no albinos previously. Why does this happen?

Although the story is rather complicated (because albinism is in the category of recessive traits), when a trait such as albinism that can be passed on to offspring crops up in a population, we say that the new trait is the result of a mutation, a change in the DNA material of the individual's gene. Mutations, although rare, are considered to be the basic source of all inherited differences in the plant and animal kingdoms. And since we are all different from each other in many inherited ways, we are all derived from mutants of one sort or another. A more straightforward example of mutation is the appearance of a white forelock or some other dominant trait cropping up in a person whose parents do not have it. Here the effect of the mutation is immediately noticeable: only one dose of the mutated gene is necessary for the forelock phenotype to appear.

One reason for the existence of albinos in populations, then, is that mutations occasionally occur in the normal pigment genes in the population. But most albino animals or people probably do not live to adulthood, and thus do not pass on their albino genes to offspring: either they are too conspicuous to predators (e.g., albino field mice are not camouflaged to a hawk's searching eye), or they are constitutionally weak (e.g., albino people can easily die of complications resulting from sunburn). Nevertheless, in some populations albinos are relatively more common than can reasonably be accounted for by rare mutations alone.

This is true of the Hopi Indians, mentioned earlier, and of several other groups that have been studied, despite the sunburn and blindness problems encountered by the albinos. Among the Hopi, albinos are common because of the effects of inbreeding in small populations (as noted earlier) and because they are sheltered. It is considered good fortune to have them around, and they are not expected to work in the Hopi corn fields under the blazing Arizona sun. Instead, albinos in Hopi-land are valued as story tellers, weavers, and the like; they have ample opportunity for sexual activity and thus reproductive success. The Hopi Indians remark that "Albinos are smart, clean, nice and pretty. . . . We are very proud of them. . . . We take good care of them."* Thus Hopi albinos have a selective advantage *socially* that outweighs the physiological defects which would otherwise put them at a selective disadvantage reproductively in the competition for gene survival.

SUMMARY

This chapter has provided a preview of the methods by which geneticists study inherited traits in humans. When the first concepts of genetics were formulated by Mendel over a hundred years ago (Chap. Two), the concern was to demonstrate that characteristics were indeed inherited from one's parents through the reproductive cells according to simple rules of gene transmission. But today we can study genes at all levels of biological organization—from the minute scrap of DNA chemical material that constitutes a single gene, to the effect of that gene on the health of individual cells, of individual people, and even of whole populations.

STUDY QUESTIONS

1. Construct a genotype diagram for generation VI, and their parents in generation V, in Figure 3–4; use AA for homozygous normal, Aa for heterozygous normal, and aa for albino.

---

*C. M. Wolff and F. C. Dukepoo: Hopi Indians, inbreeding, and albinism. Science *164*:36, 1969.

2. In Figure 3–4, what would be the expected proportions of normals and albinos among the offspring of generation III, numbers 7 and 8, if they had had a large number of children? Why is the expected phenotypic ratio different from the observed ratio?

3. Why do albinos usually have normally pigmented parents?

4. Key terms in this chapter:

pedigree	inbreeding	DNA
cell	chromosome	enzyme
		mutation

5. Suppose the mother of your friend Joe has a dominant hereditary defect. Joe doesn't exhibit the defect but is worried that he might pass it on to his children. Should he worry?

6. Long before study of family pedigrees became part of the geneticists' analytic techniques, a few people were curious about rare deformities which run in families. Before 1750 (and thus long before Mendel) the lively French scholar Maupertuis investigated such a family and described it with twentieth-century thoroughness:

> Jacob Ruhe, surgeon of Berlin, . . . was born with six digits on each hand and each foot. He inherited this pecularity from his mother, Elisabeth Ruhen, who inherited it from her mother, Elisabeth Horstmann, of Rostock. Elisabeth Ruhen transmitted it to four children of eight she had by Jean Christian Ruhe, who had nothing extraordinary about his feet or hands. Jacob Ruhe, one of these six-digited children, espoused at Dantzig in 1733, Sophie Louise de Thüngen, who had no extraordinary trait. He had by her six children; two boys were six-digited. One of them, Jacob Ernest, had six digits on the left foot and five on the right: he had on the right hand a sixth finger, which was amputated; on the left he had in the place of the sixth digit only a stump.*

(a) Diagram this pedigree. (Use diamond-shaped symbols for persons of unknown sex.)

(b) Does this rare trait (polydactyly) appear to be dominant or recessive?

(c) What kind of marriages that do not occur in this pedigree could yield useful information about dominance or recessiveness of polydactyly? How?

SUGGESTED READINGS

Baer, A. (ed.). Heredity and Society: Readings in Social Genetics. Ed. 2. New York: Macmillan Publishing Co., Inc., 1977.

Cavalli-Sforza, L. L., and W. F. Bodmer. The Genetics of Human Population. San Francisco: W. H. Freeman & Co., 1971.
An advanced text.

Dobzhansky, T. Mankind Evolving. New Haven: Yale University Press, 1962.

Lerner, I. M., and W. Libby. Heredity, Evolution, and Society. Ed. 2. San Francisco: W. H. Freeman & Co., 1976.

McKusick, V. A. Mendelian Inheritance in Man: Catalogs of Autosomal Dominant, Autosomal Recessive, and X-Linked Phenotypes. Ed. 4. Baltimore: Johns Hopkins University Press, 1975.

Stern, C. Principles of Human Genetics. Ed. 3. San Franciso: W. H. Freeman & Co., 1973.

Sutton, H. E. An Introduction to Human Genetics. Ed. 2. New York: Holt, Rinehart, & Winston, Inc., 1975.

Thompson, J. S., and M. W. Thompson. Genetics in Medicine. Ed. 2. Philadelphia: W. B. Saunders Co., 1973.

Wolff, C. M., and F. C. Dukepoo. Hopi Indians, inbreeding, and albinism. Science 164: 30–37, 1969.

---

*From B. Glass: Maupertuis and the beginning of genetics. Quarterly Review of Biology 22:201, 1947.

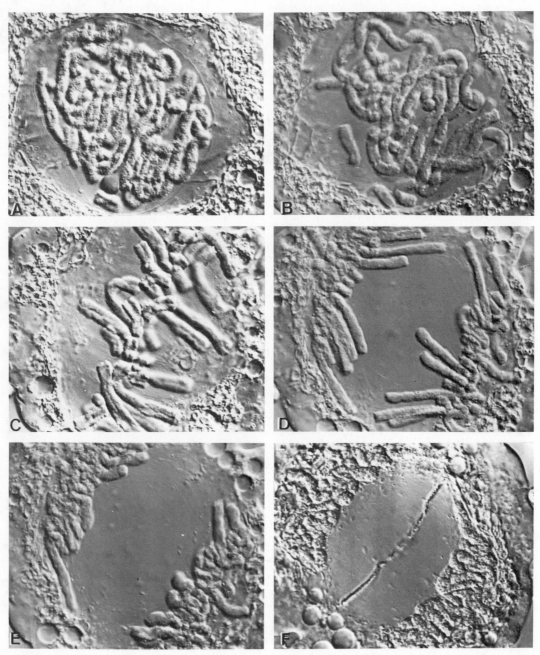

Plant mitosis. (From De Robertis, E. D. P., F. A. Saez, and E. M. F. De Robertis. Cell Biology. Ed. 6. Philadelphia: W. B. Saunders Co., 1975.)

Chapter Four

GENES ON CHROMOSOMES

THE UNIVERSAL CELL

Cells, cells, cells—we are all composed of millions of them, but bacteria, yeast, and amebas have only one cell each. In a human being all blood cells, brain cells, reproductive and other cells derive from the original, unique cell that is conceived by the biological fusion of egg and sperm. From this fertilized egg, new cells arise by cell growth and a strict program of cell division, producing a richly diverse pattern of cellular tissues and organs. It is worth our while to look closely at cells, not only because they contain the indispensable genes, but also because they embody the delicate architecture of all life on the planet earth.

Nothing man-made displays anything close to the intricate patterns in miniature of a cell seen through the microscope (Fig. 4–1). In fact, when Leeuwenhoek first looked at wriggling cells of many kinds through his peerless handmade microscope (about 1677), he was charmed with their features. A microscope is, essentially, only a clever arrangement of lenses capable of magnifying an illuminated object many times (up to 1000 × with a conventional light microscope). Although good microscopes were not developed and available to the laboratory biologist until the early twentieth century, by 1880 it was nonetheless clear—even with the bleary images of the imperfect scopes—that all body cells were formed by division of pre-existing cells, beginning with the fertilized egg (zygote).

In higher plants and animals, as opposed to bacteria and their relatives, cell division occcurs by a flow of events called mitosis. A new cell, one just formed by an old cell dividing in half, is pictured schematically in Figure 4–1. It has an outer skin, the cell membrane. Although this membrane allows minute food materials and other chemical molecules to enter, while other substances (e.g., toxic or waste matter) go out, it keeps intact the fluid interior of the cell. Within the cell membrane is the cytoplasm, a thickish substance containing a variety of special workshops, the organelles, which give the cytoplasm a fruit gelatin-like appearance. Within the cytoplasm is the nucleus, which is bounded by a membrane that selectively lets material trickle in or out through its pores.

Chromosomes. The nuclear contents in a newly formed cell are murky in appearance, like a swirling cloud when the cell is alive. This chromatin cloud, however, contains delicate fibers that become conspicuous as the cell grows and prepares to divide. During that process these fibers become the short, stubby bodies known as chromosomes—linear sequences of genes. In addition to its chromosomes, the nucleus contains a dense amorphous mass called the nucleolus, which (as we shall see later) is rich in certain gene products. Nucleoli are attached to specific chromosomes in the cell nuclei.

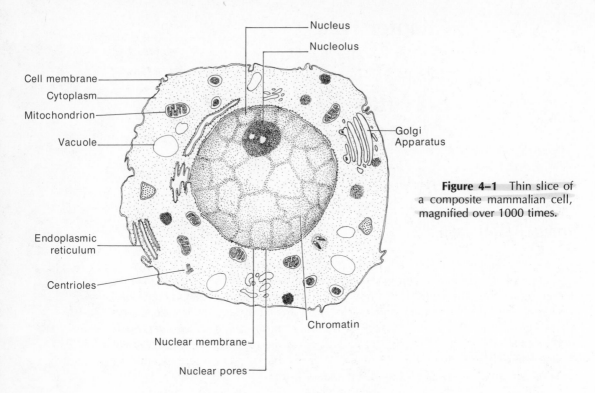

Nucleus

Nucleolus

Cell membrane

Cytoplasm

Mitochondrion

Vacuole

Golgi
Apparatus

Figure 4–1 Thin slice of a composite mammalian cell, magnified over 1000 times.

Endoplasmic
reticulum

Centrioles

Chromatin

Nuclear membrane

Nuclear pores

By patient study it is now known that different species have characteristic numbers of chromosomes. Human beings have 46, chimpanzees have 48, wheat plants have 42, 28, or 14 (depending on the species), corn plants have 20, pea plants have 14, and the fruit fly, *Drosophilia,* has 8. But as a general rule all the cells of any one individual, except for reproductive cells, have the same number of chromosomes. This constancy of chromosome number in the individual and within a whole species serves to maintain the distinctiveness of individual species and, thus, the delightful diversity of life.

Organelles. Like chromosomes, the various organelles in the cytoplasm are worlds in themselves, each having an elaborate structure and an orchestrated set of behaviors, or functions. The cytoplasmic organelles come in many shapes and sizes. The mitochondria are membrane-covered capsules which contain an enormous chemical labyrinth for extracting life-sustaining energy from food substances by breaking them down with the aid of oxygen. Mitochondria are so small (about the size of bacterial cells) that they are difficult to see with a light microscope. They are equipped with a set of numerous enzyme proteins for an organized attack on food: by a strict division of labor these enzymes degrade food substances into the smallest possible waste products, each step of the chemical degradation helping to keep the cell supplied with energy in the form of adenosine triphosphate (ATP).*

The layered sheet-like organelle in the cytoplasm is called the endoplasmic reticulum (ER). It often appears dotted with attached granules, the ribosomes. Ribosomes are vitally important to all cellular activity as factories for the manufacture of gene-ordered proteins of all kinds (enzyme proteins, toenail protein,

---

*Energy is defined as the capacity to do work; ATP, the principal energy "currency" of cells, will be discussed in later chapters of the book.

hemoglobin protein, and others). For our purposes, the endoplasmic reticulum can be visualized simply as the sculptured surface on which the ribosomes sit.

A more tubular appearing membrane system seen in the cytoplasm of many cells is the Golgi apparatus. It is conspicuous in cells that secrete quantities of proteins—for example, the cells that secrete digestive enzymes into the human stomach. The multilayered Golgi apparatus apparently packages proteins into microspheres which are then channeled to the cell exterior.

A number of other cytoplasmic organelles have been discovered, but only one demands our attention now: that is the centriole, of which there are two in each animal cell. In ways not yet understood, centrioles take part in the separation of chromosomes to daughter cells during cell division.

MITOSIS AND THE CELL CYCLE

In all plants and animals, cell division produces two cells from one. Each new cell contains a haphazard assortment of most cytoplasmic organelles and a set, by no means haphazard, of the chromosomes. Since chromosomes contain the genes necessary for all life functions, the "waltz of the chromosomes" to daughter cells, a process known as mitosis, is a precise sequence of events. Mitosis (Fig. 4–2) is marked by several distinctive phases.

Interphase.   A new cell, such as the one shown in Figure 4–1, is said to be in the interphase state. During this phase of the cell cycle the chromosomes are indistinct; the chromosomal material is then referred to as chromatin. But as the cell takes in food and grows larger, the unravelled interphase chromosomes duplicate themselves in the seclusion of the nucleus, and then gradually contract by coiling. Chromosomal replication involves the production of additional chromosomal constituents, DNA and protein (see Chap. Seven).

Prophase.   When the chromosomes have contracted just enough to become distinct, we say the cell is in prophase of mitosis. The two centrioles (Fig. 4–1), which were together in interphase, now separate to opposite ends of the cell. Motion pictures of living cells undergoing mitosis reveal that prophase is a relatively long period during which the chromosomes twist around in the nucleus while constantly contracting.

Metaphase.   When the chromosomes are fully contracted, the nuclear membrane disappears and the chromosomes come to lie on a plane in the middle of the cell, halfway between the two centriole poles. They may be guided to the midplane by the spindle fibers, which are composed of microtubules and which become apparent at this stage (metaphase) (Fig. 4–3). These spindle fibers, emanating from the centrioles, are contractile structures formed in the cytoplasm between the poles of the cell. During metaphase, some of the spindle fibers attach to the chromosomes at a point on the chromosome called the centromere; this occurs as the chromosomes move to the midplane of the cell, forming the metaphase "plate." The spindle fibers align themselves perpendicular to the metaphase plate and thus point toward the two poles of the cell. Individual metaphase chromosomes can often be identified by their lengths and by the position of their centromeres.

Anaphase and Telophase.   The separation of chromosomes into two equivalent sets is what mitosis is all about. The process approaches its culmination with anaphase. During interphase, by the time the DNA is duplicated, the chromosomes act as double structures. During prophase they each split lengthwise into two chromatids. Following metaphase, the duplicated halves of the split chromosomes separ-

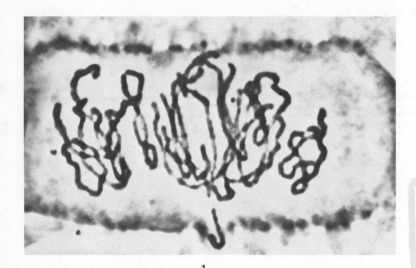

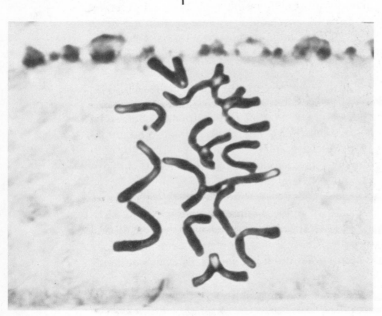

Figure 4–2 Mitosis in onion root tip cells. (1) Prophase. (2) Late prophase: the chromosomes move toward the middle of the cell. (3) Metaphase: the chromosomes are distinctly doubled. (4) Anaphase: the daughter chromosomes move toward the poles. (5) Telophase: the chromosomes become indistinct as nuclear membranes reform. (Courtesy of M. Corodemas. From Baer et al.: Central Concepts of Biology. New York, Macmillan Publishing Co., 1971.)

ate to opposite poles. Each chromatid is now seen to be attached at the centromere to its own spindle fiber and seems to be pulled by it toward the centriole at the pole on its side of the cell. Once these chromatids, or halves of chromosomes, separate from the midline, we say the cell is in anaphase. When the halves, which now are really the daughter chromosomes, reach the poles and cluster around them, we say the cell is in telophase. At this stage (in animal cells) the cytoplasm pinches itself in half at the midline of the cell. (In plants the cells lay down a plate separating the two halves.) As the chromosomes uncoil, a nuclear membrane forms around the telophase chromosome mass, thus completing the cycle and producing two new interphase cells where one existed before (see Fig. 4–4).

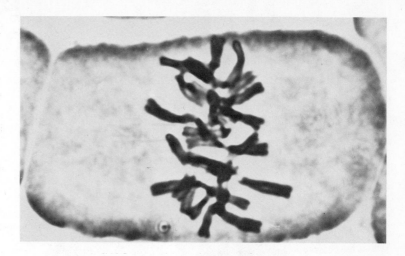

3

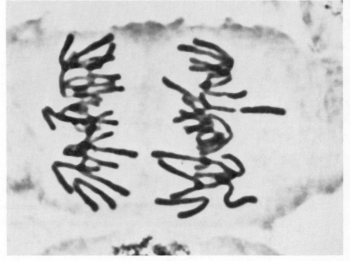

4

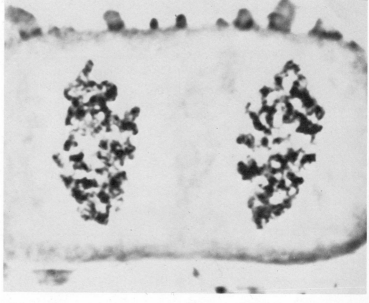

5

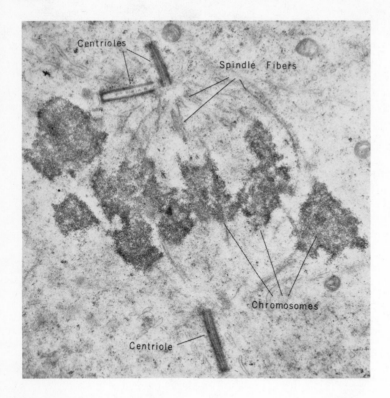

Figure 4–3 Metaphase in a mammal. (From D. W. Fawcett, *The Cell,* W. B. Saunders Co., Philadelphia, 1966.)

The Significance of Mitosis

Mitosis assures that both daughter cells get the same number and variety of chromosomes, and thus the same genetic information, as the mother cell had. That is, each duplex chromosome divides and its halves separate, one set going to each daughter cell. Rarely is there any uneven distribution of chromosomes,* or any exception to the exact partitioning of chromosomes to the daughter cells. Such deviations may have unhappy consequences, although they have taught us much about how heredity works (see Chap. Five).

The underlying rationale of mitosis is that it allows normal growth of an orga-

---

*For example, as a rare event during anaphase both halves of a #1 chromosome might go in the same direction, giving one daughter cell an extra dose of, say, the gene for Rh blood group; or both halves of a #2 chromosome might go to the other daughter cell, giving it an extra dose of, say, the gene for MN blood group (Chap. Eleven).

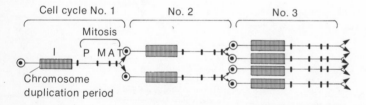

Figure 4–4 A schematic representation of the cell cycle.

I, Interphase; P, Prophase; M, Metaphase; A, Anaphase; T, Telophase

nisms' body and renewal of worn-out parts. As a child grows, his cells do not simply get bigger; they divide and produce more cells. If cells did grow larger indefinitely, they would die when they passed the critical limit beyond which oxygen and other materials could not penetrate to their centers. Oxygen is needed not just by the lungs and blood, but throughout every cell in the remotest niche of the body. It keeps the mitochondrial machinery humming for the extraction of energy from food. Moreover, a set of genes can handle only a limited amount of chemical traffic inside the cell. This limitation is demonstrated by experiments in which cell division is blocked by certain chemicals, by temperature shock, or other factors: under these circumstances the cell does grow much larger, but only at the rate that the chromosomal DNA (the gene material) increases.

By adulthood, mitosis has ceased in most human tissues, except for the replacement of old or diseased cells. Various tissues slow down and stop growing at different times. The brain slows down in infancy, the teeth later, and so on. But even in the adult some tissues continue to undergo mitosis and cell division routinely. Blood and skin are examples of these exceptional cases.

MITOSIS AND INDIVIDUALITY

Every cell in an individual's body has a full set of his particular genes carried on his set of chromosomes. Each of these chromosomes carries many genes: a human being has an unknown number of genes, at least 10,000 per cell by conservative estimates. Some of these genes are the same for all members of the species and, in fact, a subset of these is very similar among all members of all the species that constitute a recognizable, related group (such as the mammals). But other genes that an individual possesses form a unique combination for that person alone within the species—a combination that makes him the only "genetic experiment" of its kind on earth (assuming he has no identical twin).

Fortunately, this genetic material is chemically very stable. It is further protected by cellular structure and functions. The cells sequester the genes in stable sequences on chromosomes inside their nuclei. When new cells are produced, duplicate chromosomes are distributed to daughter cells. This chromosomal duplication and distribution process, called mitosis, is a universal feature of body cells in plants and animals.

HOW MITOSIS EVOLVED

By comparing simple forms of life today with the fossil remains of small life forms in ancient rocks, it is possible to imagine how some complicated structures and functions of cells arose. A plausible picture of the origin of the machinery of mitosis (spindle fibers, etc.) in primitive cells has been described by L. Margulis. The complex chromosomes of higher organisms, containing protein in addition to DNA, are thought to have arisen from the "naked" DNA of bacterial chromosomes (see Chaps. Eight and Nine). The nuclear membrane may have arisen as an inward extension of the cell membrane. The spindle fibers, centrioles, and centromeres of complex cells are envisioned as derivatives of the whip-like tails, the flagella, of a kind of bacterium. Possibly the bacterium was swallowed by an ameba-like organism and came to live on as a dependent inside the ameba. The "tail" would permit greater maneuverability for the ameba in searching for food.*

The concept of a flagellated bacterium invading an ameba in the misty past makes sense for the primitive beginnings of the mitotic process because the whip-like tail of free-living flagellates today (and also the tail of sperm cells) has a molecular architecture very similar to that of existing centrioles. In fact, microtubules grow out from a young sperm cell to form the sperm tail, a process which is structurally like the growth of spindle fibers (also microtubules) out of centrioles during mitosis.

SUMMARY

Cells are enormously complex factories for taking in raw materials from their exterior and transforming them into necessary parts for cell growth, repair, and division. They accomplish these functions by compartmentalizing their activities, which explains why they have so many kinds of specialized organelles.

The nucleus is the innermost compartment of the cell. It contains the chromosomes, the vital genetic material of cells. As a cell grows larger its chromosomes duplicate themselves and then enter into a process called mitosis. During mitosis the duplicate chromosome sets are separated to opposite ends of the cell. The whole cell then divides in the middle, producing two daughter cells with identical genetic material. By the process of mitosis, the human organism grows and develops and is able to repair damaged tissue.

Plants and animals are composed of from one to many cells. The genetic material, and thus the chromosomes, of different species differ in important respects. These differences are perpetuated from one cell division to the next by the mitotic distribution of chromosomes to daugher cells. The precision of mitotic events is thought to have evolved among early plant and animal cells following the fortuitous fusion of different types of more primitive cells.

STUDY QUESTIONS

1. Why do cells enter mitosis after they grow to a certain size, instead of growing larger indefinitely?
2. Mitotic cell division produces two cells from one. In what ways are the daughter cells identical to the mother cell?
3. In normal human liver tissue, the chromosomes of some cells duplicate and divide but the cell itself does not divide. How many chromosomes will the cell have as a result?
4. What would happen during mitosis to a chromosome that had lost its centromere? That had two centromeres?
5. Key terms:
   microscope   nuclear membrane   spindle fiber

---

*Such an invasion of one cell by another is familiar to us in terms of bacterial parasites which cause disease. But when the parasite happens to benefit the host cell, and vice versa, we call their association symbiosis. There are one-celled animals today that harbor small symbiotic plant cells inside them. The plant cell manufactures food for the benefit of the animal host and the host presumably protects the plant cell inside it.

cell

mitosis

cell membrane

cytoplasm

organelles

nucleus (pl., nuclei)

chromosome

nucleolus (pl., nucleoli)

mitochondria

endoplasmic reticulum

ribosomes

Golgi apparatus

centriole

centromere

prophase

metaphase

anaphase

telophase

symbiosis

SUGGESTED READINGS

German, J. The pattern of DNA synthesis in the chromosomes of human blood cells. Journal of Cell Biology 20:37–55, 1964.

Margulis, L. The origin of plant and animal cells. American Scientist 59:230–235, 1971.

Mazia, D. The cell cycle. Scientific American 230:53–64, 1974.

Moore, J. A. Heredity and Development. New York: Oxford University Press, 1972.

Swanson, C. P., T. Merg, and W. J. Young. Cytogenetics. Englewood Cliffs, N.J.: Prentice-Hall, Inc., 1967.

Turleau, C., and J. de Grouchy. New observations on the human and chimpanzee karyotypes. Humangenetik 20:151–157, 1973.

Human sexuality. Carved by Idel Charley, Nukuro, Micronesia, 1967. (Collection of J. Royce.)

Chapter Five

CELLS WITH A
SEX LIFE

Like cells, individual adult organisms grow to a certain size and then reproduce, although by a different kind of process. People certainly do not reproduce by dividing in half, as cells do. If people did reproduce by splitting in two, life would be exceedingly dull because every neuter offspring would be just like the now non-existent "mother." There would be nothing biologically new under the sun, except for the rare differences brought about by mutation.

What protects most species, including man, from such a drab existence is the invention of sex, which is very old. The poets of romance notwithstanding, the transcendent importance of sex is to produce a "scrambled egg" of the different genetic virtues and vices of the parents. All the hormones and love songs of human reproduction are subservient to this evolutionary process.

In this day and age of sexual candor, it seems reasonable to assume that many people understand the physiology and psychology of the human sexual experience. However, since there is more genetic significance to sexuality than most people know about, some technicalities of the process are worthy of mention here.

SEX DEVELOPMENT

Multicelled organisms are organized such that a group of cells are set aside early in embryonic development to be the germ line, the precursors of egg or sperm cells. The rest of the cells in the body are collectively called somatic cells. In the human species, and indeed in most mammals, the male's germ-line cells sit inactive in the testes until puberty and then pump hoards of mature sex cells in the seminal fluid for the rest of the individual's life, or until he becomes superannuated. The switching on of this massive sperm production is dictated by hormones. What happens to all these cells? If they are not ejaculated, they disintegrate harmlessly.

In females, however, the future eggs (about one million) all progress part way to maturity during fetal life. Then the eggs sit, arrested in their development until puberty, at which time usually one a month surges through to ripeness in the midpoint of the menstrual cycle. Again, as for the male, the whole process is controlled by hormones. Note, however, that whereas the male reproductive system is constantly ripening sex cells, the female system is decidedly economical: a 45-year-old-woman will have ripened, say, only 450 eggs throughout her reproductive period. But a male will have ripened thousands of millions of sperm. While these facts are curious in themselves, they are also suspected of being the basis of a higher rate of certain genetic mistakes (chromosomal mutations) in older women than in older men. This is because the former have, for instance, 40-year-old eggs, but the latter have sperm stored for only a few days. The basic distinction is that eggs are the products of cells which may have been inactive for longer than 40 years, whereas sperm are the products of precursor cells which, after puberty, are continuously active.

MEIOSIS

Meiosis, the process of producing reproductive cells, or gametes, in all sexual organisms, can conveniently be thought of as a specialized form of mitosis. By about 1905, meiotic events were well understood as a result of an intensive half century of microscopic study of the reproductive cells of plants, insects, and marine animals. During the late nineteenth century, as the development of microscopic techniques made it possible to count chromosomes, it became clear that offspring have the same chromosome number as their parents. Therefore, the egg and sperm which unite to form the zygote must each have half as many chromosomes as other cells in the body.

Humans, and many species of plants and other animals, have what we call a diploid number of chromosomes (double the number found in gametes); the gametes, egg and sperm, each have a haploid number (half of the diploid). Diagrammatically, then, the life cycle for a sexual species is:

1N gamete

+   ⟶ 2N zygote →→→→→→→ 2N adult ⟶ 1N gametes
       fertilization   many mitoses   meiosis
1N gamete

where 1N stands for haploid and 2N for diploid. That is, meiosis brings about 2N → 1N and is thus a process which reduces the chromosome number.

Two cell divisions are required to bring about chromosome reduction in gamete formation. The requirement for two divisions is partially explained by the fact that the chromosomes are already duplicated before the first division of meiosis starts. Thus cells are really at the 4N level of chromosomal material (DNA) when they enter prophase of the first meiotic division (meiosis I), just as when cells enter prophase of mitosis. At the end of meiosis I, then, each cell has a 2N level of chromosomal material. In the interphase before meiosis II, further chromosomal duplication is suppressed: hence the second meiotic division reduces the 2N level of chromosomal material to the 1N level, which is normal for gametes. In summary:

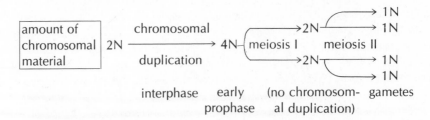

When the chromosomes going through meiosis I and II are observed through the microscope, however, it looks like the following is happening:

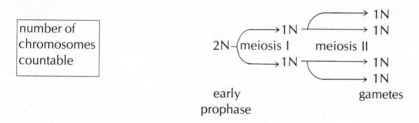

number of
chromosomes
countable

$$2N \rightarrow \begin{cases} \text{meiosis I} \rightarrow 1N \\ \rightarrow 1N \end{cases} \quad \text{meiosis II} \begin{cases} \rightarrow 1N \\ \rightarrow 1N \\ \rightarrow 1N \\ \rightarrow 1N \end{cases}$$

early
prophase

gametes

At the beginning of meiosis the chromosomes are clearly at the 2N number (46 for humans), despite the fact that they are already duplicated to the 4N level of material. At the end of meiosis I, the chromosomes are indeed at the 1N number (23 for humans) although each is still in its duplicated state, that is, each has two chromatids. Thus, the chromosome number has been reduced during meiosis I without the separation of sister chromatids, as occurs in ordinary cell divisions. But in meiosis II the chromatids *do* separate; thus, although meiosis II starts with a 1N number of chromosomes, it also ends with a 1N number—but now there are twice as many cells. Since one picture of meiosis is worth a thousand earnest words, lead yourself through the meiotic diagram (Fig. 5–1) to visualize what is happening here.

Meiosis in female mammals produces only one functional egg from every cell, or oocyte, entering meiosis. (During fetal life, mammalian eggs progress as far as

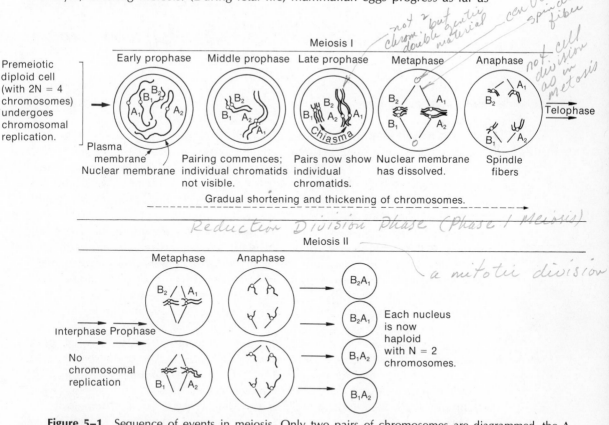

Figure 5–1 Sequence of events in meiosis. Only two pairs of chromosomes are diagrammed, the A pair being longer than the B pair. Subscripts 1 and 2 designate chromosomes derived from the mother and father, respectively. Centromeres are diagrammed as circles. (From Baer, A., Hazen, W., Jameson, D., and Sloan, W.: Central Concepts of Biology. New York: The Macmillan Publishing Company, 1976. (Copyright © 1971 by The Macmillan Publishing Company.)

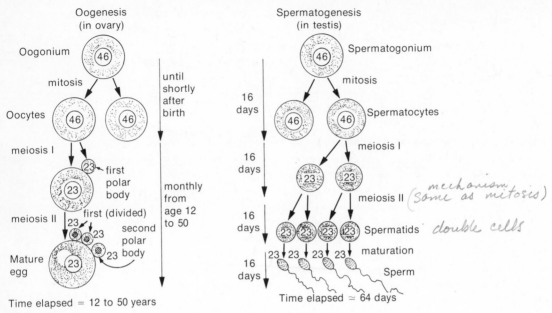

Figure 5–2   Human gametogenesis.

early prophase I.) Meiosis in males, however, produces four functional sperm from every spermatocyte (Fig. 5–2).

Chromosome Pairing and Segregation

One reason meiosis succeeds in halving the chromosome number in gametes is that, as shown in Figure 5–1, matching chromosomes pair up in prophase I. This allows members of the pair to segregate in anaphase I. The sister chromatids of each member are still stuck together (at the centromere region) at this time. It is known that the centromere region splits in meiosis II, and that sister chromatids separate only at that time.

The fact that pairing occurs in prophase I has genetic importance. Members of a pair of chromosomes are almost invariably the same size and shape—they are structurally homologous. And, as genetic breeding experiments have shown, members of a pair carry the same sequence of genes. Thus they are also homologous in a genic sense. One member of each pair comes from the mother and the other from the father. This implies that for every maternal chromosome in the egg there is a paternal counterpart, or homolog, in the sperm. And so there is.

In humans, when egg and sperm unite to form the zygote, the two sets of 23 chromosomes reconstitute the 46 chromosomes characteristic of the species, each gene in the zygote then being present in double dose. If a man and woman are both heterozygous (Aa) for the gene for albinism, located, say, on their No. 11 pair of chromosomes, then through allele and chromosome segregation in meiosis they may produce an albino child (aa). It is common speculation that, had Mendel known about segregation of homologous chromosomes (no one did in 1865), he would have happily recognized it as the physical basis of the allele segregation which he observed in his pea crosses.

Independent Segregation of Different Pairs

Other facets of meiosis rate attention in genetics. One is the fact that when, for example, chromosome pair No. 11 segregates A from a in a heterozygote and other chromosomes are also segregating their homologs with all sorts of genes on them, the particular poleward direction of the paternal homolog of pair No. 11 is independent of the direction taken by paternal and maternal homologs in other pairs. Each pair aligns itself on the metaphase plate without regard for the others. These alignments are therefore independent events. Since the chance of any one paternal chromosome going a certain direction is $\frac{1}{2}$, the chances that all the paternal chromosomes in human meiosis would go to the same pole is very small: $\frac{1}{2}$ multiplied by itself twenty-three times $\left(\frac{1}{2}\right)^{23} = \frac{1}{8,388,608}$ This means that the most common outcome for a gamete is to contain a mixture of some chromosomes derived from the mother and some from the father. By this independent behavior of chromosome pairs during meiotic reduction in chromosome number, all sorts of reshuffled sets of chromosomes are packaged into gametes. And, therefore, all sorts of gene "hands" are dealt to the next generation.

To emphasize this point, we may consider a meiotic cell with only two chromosomes pairs — one with the paternal chromosome marked A and its homolog marked a, and the second pair with the paternal chromosome marked B and its homolog b. (B and b are alleles for a second Mendelian trait.) This individual's genotype is Aa Bb, and chromosomally half his cells in meiosis will look like this at metaphase and anaphase I:

Two kinds of gametes in equal frequencies are thus produced: AB and ab. But the other half of his meiotic cells will look like this in metaphase and anaphase I:

Due to independent alignment of different pairs, two other kinds of gametes in equal frequencies are thus produced: Ab and aB.

Taking into account all the gametes produced by Aa Bb, it is clear that the processes of chromosomal segregation and independent behavior of the chromosome pairs will produce gametes that are 25% AB, 25% ab, 25% Ab, and 25% aB. (Refer back to Fig. 5-1.)

Long before these chromosomal processes were known, Mendel had anticipated their effects. In his pea plant experiments he crossed two strains which differed in two traits: this can be symbolized A vs. a and B vs. b. That is, he crossed

AA BB with aa bb, producing all double heterozygotes, Aa Bb. He found that in the cross of Aa Bb to the double recessive strain aa bb, he got four classes of offspring in equal numbers:

$$\text{Aa Bb} \times \text{aa bb} \longrightarrow \tfrac{1}{4}\,\text{Aa Bb} + \tfrac{1}{4}\,\text{Aa bb} + \tfrac{1}{4}\,\text{aa Bb} + \tfrac{1}{4}\,\text{aa bb}$$

From this 1:1:1:1 ratio he deduced that the A/a and B/b factors must segregate independently of each other to gametes—that A and B, although they were derived from the same grandparent, had no tendency to stick together in gamete formation in Aa Bb individuals. Thus, production of the 1:1:1:1 ratio resulted from random union of ab gametes from the aa bb parent with the four gamete types of the Aa Bb parent:

$$\tfrac{1}{4}\,\text{AB} + \tfrac{1}{4}\,\text{Ab} + \tfrac{1}{4}\,\text{aB} + \tfrac{1}{4}\,\text{ab} \qquad \text{(gametes of Aa Bb)}$$

$$\times \qquad \text{ab} \qquad \text{(gametes of aa bb)}$$

$$\overline{\tfrac{1}{4}\,\text{Aa Bb} + \tfrac{1}{4}\,\text{Aa bb} + \tfrac{1}{4}\,\text{aa Bb} + \tfrac{1}{4}\,\text{aa bb}} \qquad \text{(offspring)}$$

or

	AB	Ab	aB	ab
ab	Aa Bb	Aa bb	aa Bb	aa bb

Crossing Over

There is one form of reshuffling of genes in meiosis that we have not yet discussed. It is the exchange of chromosome segments between members of a pair in prophase I. This process, called crossing over, is shown diagrammatically in Figure 5–3. An X-shaped area, or chiasma, seen in a chromosome pair in prophase I indicates a crossover has taken place. Almost every cell that goes through meiosis has at least one of these crossovers, often several per chromosome pair. Crossovers can occur at any location along the gene sequence of a chromosome pair. The entwining of paired chromosomes that results from crossing over stabilizes the pair at metaphase I against nondisjunction (the abnormal condition where two homologs do not segregate and both proceed to one daughter cell [Chap. Six]).

Crossovers are a genetically enriching part of the evolutionary process. As a result of this exchange between homologs, genes that were formerly on opposite chromosomes come to be joined, or linked, on the same chromosome. Reciprocally, genes that were on the same chromosome come to lie on opposite members. For example, where gene C was formerly adjacent to gene D on one homolog and gene c adjacent to d on the other homolog, after crossing over between these two gene sites C is now adjacent to d and c is adjacent to D. Such recombination of genes scrambles the maternal and paternal contributions to each gamete even beyond the generous amount provided by independent segregation of different chromosome pairs.

A. Cytological: Observation of Paired Homologs in Meiotic Prophase I

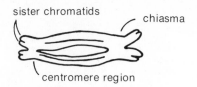

sister chromatids

chiasma

centromere region

B. Genetic

1. Meiotic Events

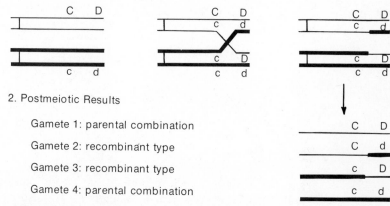

2. Postmeiotic Results

Gamete 1: parental combination

Gamete 2: recombinant type

Gamete 3: recombinant type

Gamete 4: parental combination

Figure 5–3 Cytological and genetic basis for intrachromosomal recombination. (From: Baer, A., Hazen, W., Jameson, D., and Sloan, W.: Central Concepts of Biology. New York, The Macmillan Company, 1971. Copyright © by The Macmillan Company, 1971.)

ONLY TWO SEXES

In human males and in the males of many other animal species, there is one chromosome pair whose two members are not the same size and shape. This is an exception to what was said earlier. The unlike pair consists of the so-called X and Y chromosomes.

Females have no unlike pair of homologs, but they do have a pair with members of the same size and shape as the X in the male. Overwhelming evidence points to this pair as being XX in females. As a result of chromosome segregation, then, females produce only X-bearing eggs, while males produce X-bearing and Y-bearing sperm in equal numbers.

Thus the sex ratio in "XX-XY" species is about 50:50 for female vs. male phenotypes. This is a genetic segregation ratio of 1 to 1, just as Mendel found in peas for their inherited traits when he made the cross aa × Aa. But the human sex ratio at birth (U.S. white population) is 106 males to 100 females, or a 1.06 to 1.00 ratio. Moreover, more male fetuses are known to be spontaneously aborted during the third to eighth month of pregnancy. It has therefore been surmised that the sex ratio at conception is perhaps as high as 120 males to 100 females, possibly because the smaller Y-bearing sperm travel faster in the race to fertilize the egg than do the X-bearing sperm. Recently, however, Curt Stern has pointed out that the sex ratio at conception may indeed be 1:1, with more females being miscarried during the first three months of pregnancy, followed by the excess of male miscarriages later. In U.S. whites the sex ratio goes from 1.06:1 at birth to 1:1 by age 55, and then to a higher percentage of females beyond that age as a result of males dying earlier than females.

At any rate, the genetic makeup of the sperm (the father's contribution) determines the sex of each child, old folk tales notwithstanding:

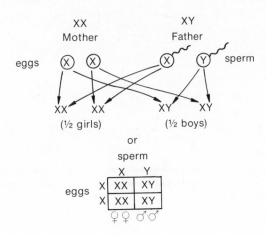

X-Linkage

The X chromosome in humans, mice, and other well-studied mammals is not only involved in male-female differences, but also carries genes that have no obvious association with sex. Red-green colorblindness and hemophilia (lack of blood-clotting ability), both of which are fairly common, are X-linked traits. (See Box on Hemophilia, p. 49.) Rare recessive traits carried on the X chromosome in humans

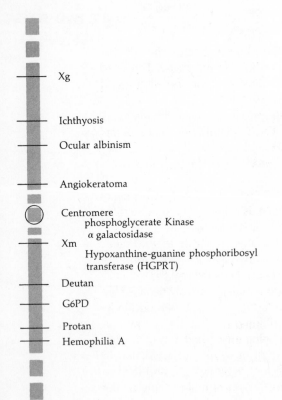

Figure 5–4 The human X chromosome. Source: W. Bodmer and L. Cavalli-Sforza. 1976, p. 68.

include agammaglobulinemia (lack of immunity to disease), some cases of cataracts and of congenital deafness, G6PD deficiency (a cause of anemia in some ethnic groups), and others (Fig. 5–4).

Recessive X-linked traits are expressed more often in males than in females. This is so because a recessive mutant allele in a female (XX) is usually "covered" by a dominant normal allele whereas the Y chromosome carries none of these genes found on the X. The segregation of the X from the Y chromosome during meiosis in males means that men pass X-linked genes to their daughters but not to their sons; sons receive only the Y from their fathers. But women who have, for example, one allele for colorblindness (heterozygote carriers) and who are phenotypically normal, pass on the colorblind allele (c) to half of their sons:

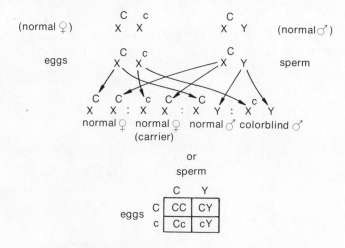

Carrier females are phenotypically normal because c is recessive to C, the color vision allele. But when a male has one dose of c he is necessarily colorblind because he has no other X chromosome to mask its effect. (A possible evolutionary value of colorblindness is that colorblind hunters can spy out color-camouflaged animals better than normal-visioned people can.)

HEMOPHILIA: THE ROYAL DISEASE

Hemophilia is an X-linked recessive defect in blood clotting. It is incurable and fatal unless treated. There are several forms. In the classical form an enzyme called Factor VIII is partially or completely inactive. Factor VIII is a vital component of the clotting process that occurs upon internal or external wounding. Hemophilia is characterized by painful bleeding in the joints due to the normal friction of muscle movement, by massive bleeding when a tooth is pulled or surgery is performed, or even by excessive bleeding due to the usual knee scrapes and bruises of childhood. Hemophilia is often detected after birth by continued bleeding from the stump of the umbilical cord or as a result of circumcision. Indeed, about A.D. 200 the Talmud recognized the problems inherent in hemophilia by prohibiting circumcision of a Jewish boy born to a woman who had previously had two sons with a "bleeder" phenotype. The prohibition applied also to the woman's sisters.

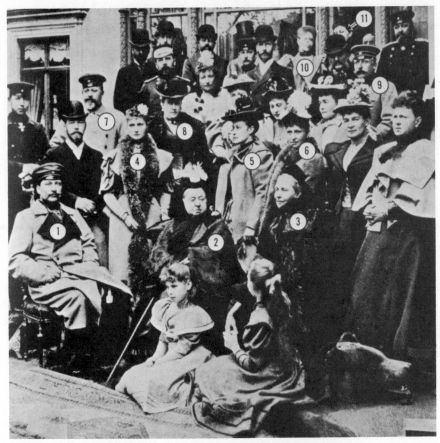

Queen Victoria of England and her descendents, photographed in 1894.

    2. Queen Victoria, a mutant carrier of hemophilia.

    3. Her daughter Victoria, Empress of Germany, a possible carrier.

    7. Her son Edward, later King of England, normal.

      (Her daughter Alice, a carrier is not shown.)

    9. Her son Arthur, normal.

    8. Her daughter Beatrice, a carrier.

    1. Her grandson Kaiser Wilhelm II of Germany, a normal son of Victoria.

    6. Her granddaughter Irene, a carrier daughter of Alice; married Prince Henry of Prussia and bore two hemophiliac sons (not shown).

      (Her grandson Frederick William, a hemophiliac son of Alice, is not shown.)

    4. Her granddaughter Alix, a carrier daughter of Alice; married Nicholas II, Tsar of Russia (standing to her right) and bore a hemophiliac heir, Alexis (not shown.)

    5. Her granddaughter Victoria Eugenia, a carrier daughter of Beatrice; married Alfonso XIII, King of Spain, and bore a hemophiliac heir.

    10. Her granddaughter Marie, a possible carrier.

    11. Her granddaughter Elizabeth, a possible carrier.

      (Her grandsons Leopold and Maurice, both hemophiliac sons of Beatrice, are not shown.)

    (From The Manseli Collection. *In* Kogan, B. A. 1970. *Health: Man in a Changing Environment.* Harcourt, Brace & World, New York.)

Within the last decade treatment has become available for hemophiliacs: Factor VIII, isolated from normal human blood plasma, is injected daily, or as needed, into the bloodstream of hemophiliacs. The current cost of this treatment is about $6,000 per year. Until recently, Factor VIII preparations, obtained from normal blood from blood banks, was sometimes contaminated with hepatitis virus. Thus hemophiliacs often incurred liver inflammation (hepatitis) as a result of their Factor VIII treatment.

There are about 20,000 hemophiliacs in the United States, virtually all males. Tests are now being developed to identify female carriers by their Factor VIII activity levels. Such tests will be a blessing for women who have hemophiliac relatives and want to know their chance of producing a hemophiliac son.

The most famous mother of a hemophiliac was Queen Victoria of England (1819–1901). She was a carrier, probably having received a fresh mutation for hemophilia from one of her parents. One of her sons was a hemophiliac and two of her daughters were carriers. This fact, combined with the intermarriage customs for European royal families, contributed to the shakiness of royal rule throughout Europe, a rule already endangered by populist opposition to kinghood. Moreover, at least one of Queen Victoria's granddaughters, who was the wife of Nicholas II of Russia and the mother of the Tsar-apparent, Alexis, was so guilt ridden about her son's hemophilia (knowing it was transmitted through her) that she unintentionally contributed to the triggering of the Russian Revolution of 1917 by accepting the medical aid of the unscrupulous court intriguer Rasputin. However, the existing royal family of Britain, which has weathered many crises, did not happen to inherit the disease. Queen Elizabeth II is a great-great-grandchild of Queen Victoria, and her husband, Prince Philip, is likewise descended from Victoria through a German branch of the family.

Colorblind females are extremely rare because they require a colorblind father (cY) and a carrier mother (Cc), and the chance of such a union is low. (In the U.S. white population only about 8 percent of men are colorblind and only about 15 percent of women are carriers; in addition, it is not likely that sharing a colorblind gene would increase marriage tendencies between two people.) The child of a carrier woman married to a colorblind man has a 1 in 4 chance (25%) of being both colorblind and female, since the expected offspring ratio is $\frac{1}{4}$ carrier ♀: $\frac{1}{4}$ colorblind ♀: $\frac{1}{4}$ normal ♂: $\frac{1}{4}$ colorblind ♂, or:

$$Cc\ ♀\quad ×\quad cY♂$$
$$\downarrow$$

sperm

		c	Y
	C	Cc	CY
eggs			
	c	cc	cY

As is the case for colorblindness, so it is for all X-linked recessive traits: males are much more often affected phenotypically than females (Fig. 5–5). Since many of

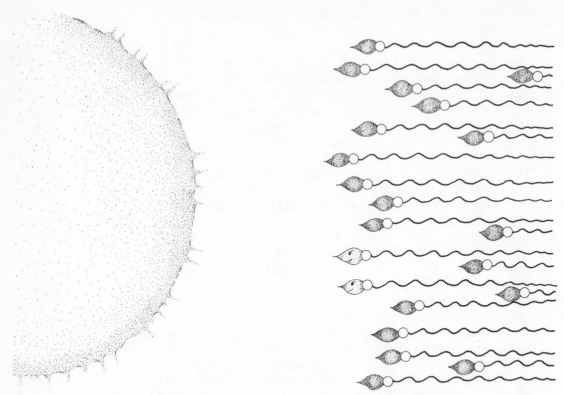

"At this point, why worry if it's got a colorblind gene or not?"

these traits are disadvantageous, X-linkage brings about a greater selective stress on males than on females. This is suspected to be one reason for the poorer survival of males from conception onward, and thus offers one explanation for the gradual change throughout the lifespan of the sex ratio among humans from an initially higher proportion of males to a preponderance of females.

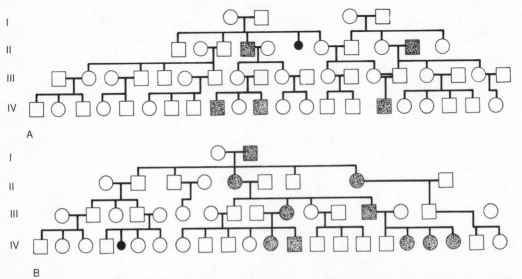

Figure 5–5   A, A not uncommon pedigree for a rare X-linked recessive trait, such as red-green colorblindness. B, In contrast, the pedigree for a rare X-linked dominant trait.

The intricate workings of meiosis faithfully produce eggs or sperm cells which have half the genetic material of somatic cells and which have different combinations of the genes present in the individual. These diverse combinations arise by chromosomal (and thus gene) segregation and independent assortment of different chromosome pairs, augmented by crossing over. The unusual nature of the X and Y chromosomes in humans produces a distinctive difference in genetic makeup between the sexes. The location of certain genes on the X chromosome (X-linkage) raises questions about their evolutionary significance.

STUDY QUESTIONS

1. What indications exist that genes are carried on chromosomes?
2. Why are people genetically different?
3. Suppose a farmer in Europe has a vigorous, dwarf variety of pea plants that has unappetizing yellow seeds. His neighbor cultivates a variety of peas that has tasty green seeds but is spindly and tall. Suppose further that the two farmers know from long observation, or from reading a genetics book, that unappetizing yellow is dominant to tasty green and tall is dominant to dwarf. (See Chap. Two and the Study Questions therein.) The farmers decide to hybridize their two varieties to produce a short, vigorous variety that has tasty green seeds. If their varieties are pure-breeding, they can then perform the cross: dwarf yellow (tt YY) × tall green (TT yy), and the hybrid (F_1) offspring will be tall yellow. (Here Y refers to the yellow allele, not the Y chromosome.)
   (a) What is the genotype of the F_1?
   (b) What genetic cross should they make the next growing season to produce a pure-breeding short (vigorous) and green (tasty) variety? [Remember that Tt × Tt gives 3 tall (T–) to 1 short (tt) and Yy × Yy gives 3 yellow (Y–) to 1 green (yy).]
   (c) Assuming that the height and the color genes are on different pairs of chromosomes (Mendel showed long ago that they did, in fact, segregate independently), what is the expected proportion of short green offspring in the cross Tt Yy × Tt Yy?
   To answer this question, set up a Punnett square as follows:

(♂ gametes)

(♀ gametes)	$\frac{1}{4}$ TY	$\frac{1}{4}$ Ty	$\frac{1}{4}$ tY	$\frac{1}{4}$ ty
$\frac{1}{4}$ TY				
$\frac{1}{4}$ Ty				
$\frac{1}{4}$ tY				
$\frac{1}{4}$ ty				

4. In the hypothetical cross Aa bb × aa Bb, where the two genes are on different chromosome pairs, what are the kinds and proportions of gametes that each parent produces? What are the kinds and proportions of offspring produced from this cross?

5. In Figure 5-5, did the third colorblind boy in generation IV receive his color-blind gene from his colorblind grandfather?

6. What is the chance that a colorblind woman will have a colorblind son? A color-blind daughter? Does either of your answers depend on the genotype of the father? How?

7. What is the most likely genetic basis of the trait shown in the following pedigree?

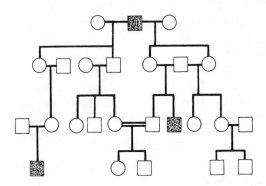

8. Key terms:

germ cells	chromosome segregation
somatic cells	independent chromosome segregation
gamete	crossing over
meiosis	chiasma (pl., chiasmata)
diploid	linkage
haploid	X-linkage
oocyte	colorblindness
spermatocyte	hemophilia
homolog	

SUGGESTED READINGS

German, J. Studying human chromosomes today. American Scientist 58:182–201, 1970.

Goss, S. J., and H. Harris. New method for mapping genes in human chromosomes. Nature 255:680–684, 1975.

Marx, J. L. Hemophilia: new information about the "royal disease." Science 188:41–42, 1975.

Moore, J. A. Heredity and Development. Ed. 2. New York: Oxford University Press, 1972.

Moses, M. J., S. J. Counce, and D. F. Paulson. Synaptonemal complex complement in man in spread of spermatocytes, with details of the sex chromosome pair. Science 187: 363–365, 1975.

People Magazine. A "fix" of life for a hemophiliac. July 22, 1974, pp. 58–59.

Ratnoff, O. D., and B. Bennett. The genetics of hereditary disorders of blood coagulation. Science *179*:1291–1298, 1973.

Ruddle, F. H., and R. S. Kucherlapati. Hybrid cells and human genes. Scientific American *231*:36–44, 1974.
  Describes recent results in identifying genes on particular chromosomes by studying cells formed by the experimental fusion of human cells with those of other animals.

Stern, C. Principles of Human Genetics. Ed. 3. San Francisco: W. H. Freeman & Co., 1973.
  Pages 529–539 discuss the sex ratio extensively.

Stern, C. High points in human genetics. American Biology Teacher *32*:144–149, 1975.

Swanson, C. P., T. Merz, and W. J. Young. Cytogenetics. Englewood Cliffs, N.J.: Prentice-Hall, Inc., 1967.

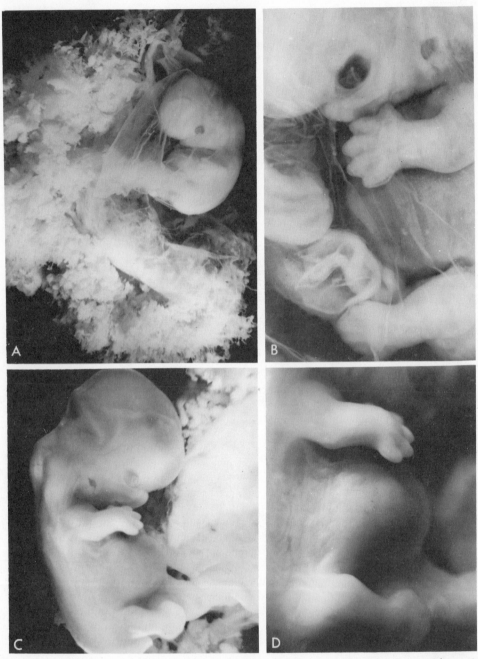

A: embryo, six weeks. B: embryo, 38 days. C: slightly older six-week embryo. D: embryo, 39 days. (From Moore, K.: The Developing Human. Philadelphia, W. B. Saunders, 1973. Courtesy of Brenda D. Bell.)

Chapter Six

CHROMOSOMES, SEX, AND CHROMOSOME ABNORMALITIES

William Bateson, the English geneticist who popularized Mendel's achievements after his work was rediscovered in 1900, once said, "Treasure your exceptions." In genetics this is excellent advice because many riches have been unearthed by some discerning mind contrasting a new mutation or recessive trait with the normal condition. By these comparisons we come to understand what genes do.

SEX DETERMINATION AND SEX DEFECTS

Human sex determination is a case in point. Simply knowing that females are XX and males are XY, as discussed in the previous chapter, does not tell anyone if the X and Y determine membership in a sex, or if other factors cause sex differences while the X and Y just follow along passively. Are the X and Y the cause of sex or its effect?

Some recently discovered exceptional mistakes in sexual development show that the Y chromosome determines human maleness and that lack of the Y determines femaleness. The story goes back about 25 years, when Murray Barr, a Canadian anatomist, unexpectedly found that females have a dark chromosomal spot in the nuclei of many of their cells, but that cells from males do not have this spot (Fig. 6–1). When one scrapes off some flat (dead) cells from the inside of the cheek, spreads them onto a microscope slide, stains them with a drop of red dye, and looks at them through the microscope, many of the cells show this distinctive nuclear spot, called a Barr body, if they come from a normal female. None show the spot if they come from a normal male.

Following Barr's work, later investigators found that a few human females who were regarded medically as having Turner's syndrome, that is, as being subnormal in their sexual development, had no Barr body. Other rare females had two or more Barr bodies. A few males with one or more Barr bodies were also found, usually males who had been diagnosed as having Klinefelter's syndrome, another abnormal-

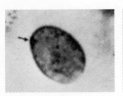
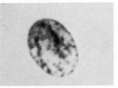

Figure 6–1 Sex chromatin in human cells. Note the Barr body *(arrow)* in the female cell on the left, and the absence of it in the male cell in the middle. Right: XXX cell. (From M. L. Barr: Cytologic tests of chromosomal sex. Progress in Gynecology *3*: 131–141, 1957. By permission, Grune & Stratton, Inc., New York.)

ity in sexual development. (A syndrome is a characteristic group of symptoms that occur together.) All these findings lacked explanation until cytogeneticists (specialists in chromosome studies) began diligently characterizing chromosomes of people with severe congenital defects. This happened only after techniques had been developed for counting and identifying human chromosomes, a process known as karyotype analysis (Fig. 6–2). The technology of karyotyping is discussed later in the chapter.

Severe congenital defects were canvassed for chromosomal abnormalities on the hunch that if an individual had too few or too many chromosomes, or even just a piece of one missing, so many genes would be affected that the person might be conspicuously abnormal. As expected, some severe abnormalities did turn out to have abnormal karyotypes (Tables 6–1 and 6–2). Among the sexual abnormalities, females with Turner's syndrome (ovary defects, sterility, and other problems) turned out to have 45 chromosomes instead of the 46 found in normal females and males (Fig. 6–3). Since the missing chromosome in Turner cases was shown to be one of the X's, Turner females are 2N = 45 and "XO." The fact that Turner females lack one X and also lack a Barr body suggested that the nuclear spot found in XX females is probably an X chromosome. (The Lyon hypothesis to explain this spot is discussed in Chap. Nine.)

Something was also found to be amiss in a few males: some had a karyotype of 47 chromosomes, the extra one being either an X or a Y. The 47,XXY males have a Barr body and are classified medically as having Klinefelter's syndrome (Fig. 6–4), which is characterized by poor testicular development and sterility (among other

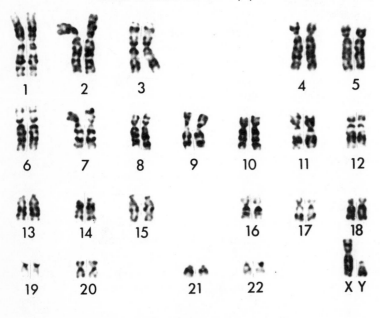

Figure 6–2 Normal human male karyotype showing banding pattern resulting from Giemsa stain. Each chromosome can be identified. (From Thompson, J. S., and M. W. Thompson. Genetics in Medicine. Philadelphia: W. B. Saunders Co., 1973, p. 16. Photomicrograph courtesy of R. G. Worton.)

TABLE 6–1   Human Sex Chromosome Abnormalities

Phenotype	Sex Chromosome Condition	Number of Barr Bodies	2N	Fertility
Normal females	XX	1	46	Yes
Turner's syndrome females	XO	0	45	No
Triple X females	XXX	2	47	Some
Tetra X females	XXXX	3	48	Yes
Penta X females	XXXXX	4	49	Does not survive
Normal males	XY	0	46	Yes
Klinefelter's syndrome males	XXY	1	47	No
Double Y males	XYY	0	47	Yes
Triple X males	XXXY	2	48	No
Tetra X males	XXXXY	3	49	No

things). On the other hand, the 47,XYY males have no Barr body and are not recognized medically as having any syndrome. XYY's are fertile and apparently healthy, though unusually tall. The theory that all XYY's have aggressive tendencies is not confirmed by current evidence (see Chap. Fourteen).

These findings suggest that the number of Barr bodies is always one fewer than the number of X chromosomes in a cell. More recent studies of individuals with XXX, XXXY, and other extraordinary numbers of sex chromosomes (whether extra X's or Y's) bear out this rule on Barr bodies (Table 6–1). Except for XYY, the life span of people with these abnormal chromosome conditions is probably somewhat

TABLE 6–2   Characteristics of the Commonest Chromosome Abnormalities in the U.S.*†

Condition	Estimated Incidence At Birth	Calculated Number Born Per Year	Probability of Survival Once Born	Fertility	Other Characteristics
Turner's XO	1 in 10,000	500	Probably almost normal	No	Sexual infantilism, webbed neck, short stature, rarely mentally retarded
XXX	1 in 1600	3100	Probably almost normal	Some	Irregular menstruation, some mental retardation; mental illness common
Klinefelter's XXY	1 in 800	6200	Probably almost normal	No	Sparse body hair, some breast growth, small testes and penis; generally mentally retarded
XYY	1 in 1000	5000	Probably normal	Yes	Tall stature, some mental retardation
Down's syndrome (trisomy 21)	1 in 700	7100	Some reach adulthood	♀♀, some ♂♂, no	Mental retardation (17% of U.S. total), deformities, leukemia, heart trouble, and respiratory infections common
Trisomy D	1 in 10,000	500	0–3 years	—	Many deformities
Trisomy E	1 in 4000	1200	0–4 years	—	Many deformities

*Assuming a population of 220 million with a birth rate of 5 million per year.
†Modified from V. A. McKusick: Human Genetics. Ed. 2. Englewood Cliffs, N.J.: Prentice Hall, Inc., 1969, p. 31.

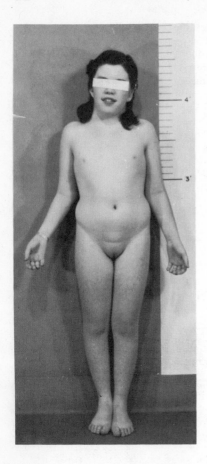

Figure 6–3 Female with Turner's syndrome (XO). Such individuals are usually short, have broad (webbed) necks, and undeveloped internal and external sexual characteristics. They are sterile. Their intellectual development is usually within the normal range. Most fetuses with Turner's syndrome die long before the end of the pregnancy period; thus there are fewer Turner females born alive than Klinefelter males. (From M. L. Barr: Sexual dimorphism in interphase nuclei. American Journal of Human Genetics 12:118, 127, 1960.)

shortened (Table 6–2); drastically deviant chromosome conditions have been found only in spontaneously aborted fetuses, never in live infants.

Where does this take us? The question whether the X or Y chromosome causes differences between the sexes is now answered. Because XO, XX, and XXX individuals are phenotypically female while XY's, XYY's and XXY's are phenotypically male, the presence of the Y chromosome in a person determines his body to be male, although not necessarily fertile. The presence or absence of a Y determines the most sublime phenotypic difference recognized in human society—male versus female.

Early human embryos appear undifferentiated, or neutral, in sexuality until about 8 to 10 weeks after conception (Fig. 6–5). The sex chromosomes trigger the development of internal and external sexuality. Under the influence of the Y chromosome, the gonads in an XY embryo begin to secrete testosterone (male sex hormone) during the sixth week of development. Testosterone induces the gonads to develop into the testes, induces the fetal female (Mullerian) ducts to degenerate and induces the fetal male (Wolffian) ducts to develop into the vas deferens (sperm ducts). At the eighth month the testes descend into the scrotal sac. In an XX embryo the gonads produce estrogen (female sex hormone) by the third month after conception and thereupon are transformed into ovaries. At that time the Mullerian ducts become oviducts (and uterus) and the Wolffian ducts degenerate. In an XY embryo the rudimentary external genitalia become the penis and scrotum; in an XX embryo, the clitoris and labia. Both sexes have both male and female sex hormones, males

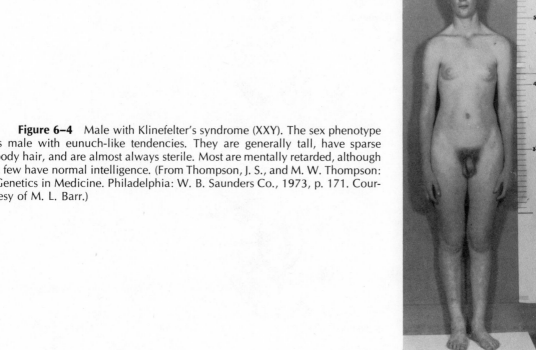

Figure 6–4   Male with Klinefelter's syndrome (XXY). The sex phenotype is male with eunuch-like tendencies. They are generally tall, have sparse body hair, and are almost always sterile. Most are mentally retarded, although a few have normal intelligence. (From Thompson, J. S., and M. W. Thompson: Genetics in Medicine. Philadelphia: W. B. Saunders Co., 1973, p. 171. Courtesy of M. L. Barr.)

simply having much more testosterone and females having much more estrogen (and other female hormones).

Causes of Sex Chromosome Abnormalities

The researchers in human genetics who first found individuals with XO (45) and XXY (47) chromosome constitutions realized that genetic studies on laboratory animals, such as the fruit fly *Drosophilia,* had long before shown how these abnormalities could be caused by mistakes occurring in meiosis.

In many organisms the following kind of mistake occurs as a rare meiotic event: a pair of homologous chromosomes fails to segregate, as they should, in the first meiotic division, and both homologs go to the same pole of the cell. This nonsegregation of homologs, or nondisjunction as it is called, means that one daughter cell has an extra chromosome, and the other daughter cell is missing one. These daughter cells go through the second meiotic division without any mistakes, thus producing the usual total of four cells from the one "starter" cell. But these four cells have unequal numbers of chromosomes: two of the four cells have 22 chromosomes each and two have 24 (Fig. 6–6). If a sperm cell with 22 chromosomes fertilizes a normal egg (23 chromosomes), the offspring will have 45 chromosomes. If the chromosome missing from the sperm cell was a sex chromosome, the offspring would be XO and show Turner's syndrome. The same argument can be made for a sperm with 24 chromosomes that contains both an X and a Y; when such a sperm fertilizes a normal egg, the result is 47 chromosomes and the XXY condition. An XX-bearing

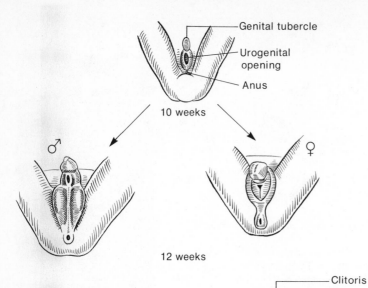

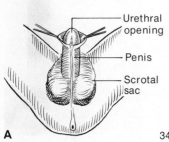

Figure 6–5 Sexual differentiation in early human embryonic development. *A,* The external genitalia develop from a common embryonic stage. *B,* The internal sex organs develop from a common embryonic stage.

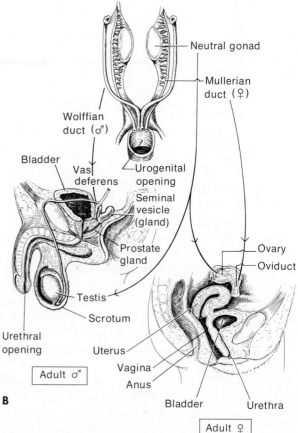

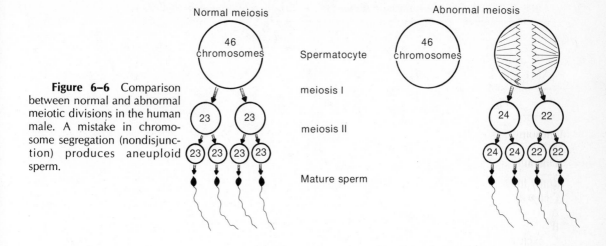

Figure 6-6 Comparison between normal and abnormal meiotic divisions in the human male. A mistake in chromosome segregation (nondisjunction) produces aneuploid sperm.

sperm can lead to an XXX female, and a YY-bearing sperm can lead to an XYY male. Both these latter abnormalities can arise from nondisjunction of the sex chromosomes in meiosis II, but not from nondisjunction in meiosis I.

Can the Klinefelter karyotype of 47 chromosomes with XXY be produced by abnormal segregation of chromosomes in the mother? Yes, if an unfertilized egg has two X's due to nondisjunction of the X pair in meiosis. A 24-chromosome (XX) egg fertilized by a normal, Y-bearing sperm (23 chromosomes) would give a 47,XXY individual. The corollary to this XXY type would be an XXX female from an XX-bearing egg and a normal X-bearing sperm. A 22-chromosome egg lacking both X's would produce XO or YO upon fertilization. A YO condition must be immediately fatal—it has never been detected.

If these mechanisms were the only ones producing XO, XYY, XXX, and XXY human beings, and if all these abnormalities had equal viability (survival), we would expect XO and XXY to be more common than, say, XYY because they result from nondisjunction mistakes in both males and females. But, as shown in Table 6-2, XYY is much more common than XO. Survival is known to be poorer for XO embryos (though apparently not for XXY or XYY) than for those with normal diploid constitutions. Studies on spontaneously aborted fetuses have found many (25 to 50% in most samples) with abnormal chromosome numbers (Table 6-3). In any event, these chromosome abnormalities (and others), despite their relative rarity, do occur repeatedly. Indeed, they occur often enough to be a public health problem in the U.S. (Table 6-2).

Adults with XXX or XYY who reproduce would be expected to produce further "health problems" in the form of XXY, XXX, and XYY offspring, but this has not been found. XYY men have produced only chromosomally normal offspring; XXX women, in one survey, produced 29 normal and only one XXY offspring. Either there is strong selection against chromosomally abnormal offspring, or some other process, as yet undiscovered, is operating.

HOMOSEXUALS, TRANSSEXUALS, HERMAPHRODITES

There is no known relationship between chromosome abnormalities and homosexuality. But people in the U.S. with sex chromosome abnormalities probably are less likely to be homosexual than people with normal karyotypes, simply because

TABLE 6–3   Types of Chromosome Abnormalities Found in Spontaneous Abortions*

	No. Found
Monosomy (2N = 45)	
XO	12
One autosome missing	1
Trisomy in various chromosome groups (2N = 47):	
A	2
B	1
C, excluding the X	6
XXY	1
D	11
E	20
F	0
G	9
Double trisomy, XXY + 1 autosome (2N = 48):	1
Triploidy:	
69	8
69 + 1 autosome	2
Tetraploidy:	
92	3
92 + 2 autosomes	2 (twins)
Other abnormalities:	3
	82 of 152 tested (54% abnormal)

*After T. Kajii et al., 1973.

the former usually have enough developmental problems of their own. This subject, however, has not been thoroughly analyzed, in part because homosexuals are not well accepted socially and thus are not well identified in our society.

Transsexuals are usually males who have changed phenotypic sex by surgical or hormonal treatment. The change is usually done for psychological reasons. These so-called sex reversals are often desirable because of ambiguity in external genitalia at birth.

The word "hermaphrodite" comes from Greek mythology (Hermes, a god + Aphrodite, a goddess). It means having attributes of both sexes, some partial development of both ovaries and testes and of the corresponding secondary sexual traits (breast development, beard). Despite the sex hormone imbalances underlying the

TABLE 6–4   Sex Phenotype Determinants and Components*

Level of Organization	Variables
1. Genetic sex	Chromosomal
2. Hormonal sex	Proper androgen/estrogen balance; different for females and males
3. Internal sex	Ovaries, testes, or mixture; reproductive ducts (different in females and males)
4. External sex	Genitalia, hairiness, breasts
5. Psychological sex	One's social personality

*Modified from J. Money. Sex Research: New Developments. New York: Holt, Rinehart, & Winston, Inc., 1965.

development of human hermaphrodites, most of them are chromosomally normal. The cause of these rare double-sex phenotypes, or of others of ambiguous sex (intersexes), is unknown for the most part: a few hermaphrodites are XX, XY mosaics (discussed later in the chapter). At any rate, no human being is known who has both functional egg and sperm cells, and the same is true of other mammals.

One's sex phenotype has many levels of organization (Table 6–4), any of which can "go wrong." The predominance of sexual normality suggests that sex development is controlled by a marvelously efficient regulatory process in the body.

OTHER ABNORMALITIES IN CHROMOSOME NUMBER

Two main kinds of chromosome abnormalities afflict humanity and other groups, including some plants and animals very important to human survival. One kind of abnormality is the presence of an extra set or sets of chromosomes, a condition known as polyploidy [*poly* (many) + *ploid* (set)]. The other kind, involving gain or loss of individual chromosomes (such as we have already discussed for the X's and Y's in humans), is referred to as aneuploidy [*an* (away from) + *eu* (true) + *ploid* (set)]; that is, aneuploidy means differences of other than a full set of chromosomes. (Diploidy differs from haploidy by a full set of chromosomes; this is a euploid, not an aneuploid, difference.)

Chromosomes are numbered in order of decreasing size, and human chromosomes are, in addition, subdivided into groups according to similarities of centromere position. No newborn child has yet been found with extra, or missing, chromosomes of the largest size (groups A and B), presumably because the large chromosomes harbor such an enormous number of genes that deviation in their number unbalances very many cellular functions—to the extent that life cannot exist. But the smaller human chromosomes (groups C through G) have been found in odd numbers in live offspring. Even then, the chromosomal upset has devastating consequences: these children usually do not survive to adulthood (Table 6–2).

Aneuploidy

The most common aneuploid upset in human beings that does not involve the X or Y chromosomes is a chronic disease called Down's syndrome or trisomy 21. Trisomy means having three copies of a chromosome instead of the normal two copies. Trisomy 21 refers to the 21st of the 23 chromosomes in the human set. Because of its smallness, chromosome 21 is put in the G group of chromosomes. Down's syndrome people (erroneously called mongoloids formerly) are characterized by mental retardation, growth defects, and many other defects throughout the body. They rarely live to adulthood.

Children with Down's usually occur sporadically in the population. It is rare for a family to have two such children (and when they do it indicates a more complex chromosomal situation, as discussed later in this chapter). Since trisomy 21 is due to an accident in meiosis, its sporadic occurrence is not surprising. The curious thing is that this accident happens more often in older women (say, over 35) than in younger women or in men (Fig. 6–7). That is, women over 35 have a much higher than usual chance of having a Down's syndrome child. (Thus the trend toward women having all their children in their 20's in technologically advanced countries is genetically beneficial.) Probably more accidents of meiosis happen in older women

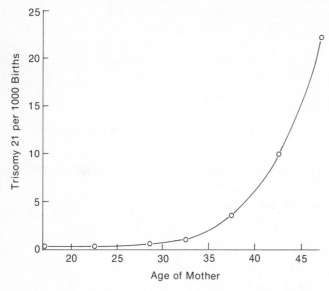

Figure 6–7 Increase in risk of child with Down's syndrome as mother's age increases.

than in older men because an older woman's egg population is correspondingly old. Remember that women produce no new eggs after birth, whereas men are constantly producing new sperm cells during their reproductive years (Chap. Five). It may also be that sperm with chromosome abnormalities are not very lively, that they are poor competitors in the race toward fertilization.

Polyploidy

Just as aneuploid accidents befall people as well as many plants and animals, human beings and other organisms are occasionally found with polyploidy, that is, with more than two sets of chromosomes. Three sets of chromosomes is called triploidy (3N), four sets is called tetraploidy (4N), and so on.

In humans, polyploidy is very rare and virtually incompatible with life. The few triploid (69 chromosomes) or tetraploid (92) cases reported have been either of aborted fetuses or of infants who died soon after birth. Again, the best explanation for the fatality of these chromosomal upsets is that they somehow enormously unbalance the complex functions of cells. Laboratory studies on plants and animals help us to understand how polyploidy can come about: chromosomes may duplicate and separate into daughter chromosomes correctly, but then fail to move away from the cell midline in mitotic metaphase; the cell does not divide in two, the nuclear membrane encloses the single cluster of chromosomes, and the cell thus contains a doubled number of chromosomes. A diploid (2N) human zygote could suffer this failure of mitotic division and produce a tetraploid (4N) cell. If this failure happens in the first division of meiosis, a sperm or egg cell will be produced which is diploid (2N) instead of haploid (1N). Normal fertilization of a 2N egg produces a triploid (3N) zygote. (Note that triploidy refers to three whole sets of chromosomes, whereas trisomy refers to three copies of only one kind of chromosome.)

Triploid organisms are almost always sterile. Cultivated bananas are triploid and sterile; that is why they are seedless. In contrast to triploidy, which is rare among plants and animals, tetraploidy is not uncommon, particularly among plants that people find useful in one way or another. Many of our cultivated crops are tetraploid or have even higher levels of polyploidy, unlike their scrawnier but tougher wild (and diploid) relatives. The cultivated cotton plant is 4N, wheat is either 4N or 6N, tobacco is 4N, and the potato is 4N. How plants can tolerate polyploidy while animals generally cannot is not known; but polyploidy in plants may be related to their poorly understood mechanisms of sex determination (different from those of animals).

Mosaics

If you look at one tile in a mosaic floor you cannot see the whole pattern. By analogy, the diagnosis of an individual's chromosomal makeup solely by observing the number of chromosomes in a few dividing cells from a blood sample does not show the whole picture. More careful studies, using a variety of tissues and analyzing a larger number of cells, make it clear that loss of laggard chromosomes during mitotic division, as well as nonsegregation mistakes that produce aneuploidy and polyploidy, occurs occasionally during the ordinary development of various tissues in the body. Not all XO females or Down's children have chromosome abnormalities in every cell of their body; some have their abnormality only in certain tissues and cell types and are chromosomally normal in other cells. Such "mosaic" individuals may have only mild symptoms of their diagnosed syndrome. On the other hand, normal people occasionally have a few cells with one chromosome abnormality or another (curiously, liver cells are regularly polyploid), and older people have more of these deviant cells than young people do.

ABNORMALITIES WITHIN CHROMOSOMES

In about 4 per cent of the Down's cases known the extra chromosome 21 is found to be stuck onto a normal chromosome number 15 (or some other chromosome). The attachment of all or part of one chromosome to another that is not homologous with it is called a translocation (Fig. 6–8). This piggyback, or translocation, form of Down's syndrome is curious because in such instances one of the parents, although phenotypically normal, often has the same rearranged chromosome as the Down's child. The chromosomally abnormal parent bears a balanced translocation: the 15/21 chromosome is "balanced" by the presence of the reciprocal 21/15 chromosome. Unfortunately, the parent with a balanced translocation can produce a gamete that contains one normal 21 chromosome plus the 15/21 translocation chromosome. Upon fertilization by a normal gamete (one copy of 15 plus one of 21), the product of this union contains two 15's plus almost three copies of 21. Such an offspring has Down's syndrome symptoms. The translocation parent who thinks Down's children result only from sporadic accidents may unwittingly produce several Down's children from his or her "harmless" translocation. Explanation of the situation by a genetic counselor can often preclude such problems (see Chap. Fifteen).

Other kinds of structural abnormalities of chromosomes are known in plants and animals. Studies on laboratory organisms have shown how these abnormalities can arise and how they can be passed on to descendants. A piece of a chromosome can be detached and disintegrate (not reattach elsewhere), thereby producing a deletion of some genes. Quite likely the Y chromosome in humans and their mammalian ancestors was once larger than it is now, and some genes were deleted (or translocated) as it became more specialized for sexual patterning in male development. Some other instances of deletions are known in humans: for example, a deletion in chromosome 13 is frequently associated with the eye cancer known as retinoblastoma; and a deletion in chromosome 5 produces a fatal disease called cri-du-chat (catlike crying occurs in affected infants). Deletion, or loss, of the whole chromosome is of course the case in Turner's syndrome.

Duplication of part of a chromosome can also occur in many species, but no large duplications are known in humans except for Down's syndrome and others in which the duplicated piece arises through a history of translocation. Small duplications appear to have occurred in the case of numerous genes, such as those determining hemoglobin structure (Chap. Ten). Extra (duplicate) whole chromosomes occur in XXX and other human trisomies.

Transpositions of a chromosomal chunk from one part of a chromosome to another can change the gene sequence from, say, ABCDEFG to ABEFCDG or some other order. If a chromosomal chunk gets turned around in a chromosome, ABCDEFG $\longrightarrow$ ABCEDFG, we refer to it as an inversion. Inversions, translocations, and deletions have been found in human chromosomes subjected to certain drugs, x-rays, and viruses, and in cancerous cells (see Chap. Twelve). The recent development of techniques to analyze banding patterns along the length of chromosomes, using special dyes, are beginning to reveal subtle changes in chromosome structure associated with a few disease categories.

CELL CULTURE AND AMNIOCENTESIS

One problem in chromosomal analysis, or karyotyping, in many animals, and particularly humans, is that most cells in the adult body are not dividing. Many organs, such as the brain and lungs, have very little cell division after birth. Human tissues with a high mitotic rate include skin, bone marrow (where blood cells mature), the lining of the gut, and, in males, the testes. Unless an individual is willing to part with some skin, gut, or the like, how then can karyotyping be done?

Fortunately, techniques have been developed that induce white blood cells (lymphocytes) to divide in a rich food medium in laboratory flasks (in vitro). A small blood sample is taken from a person and is cultured in the medium, to which has been added a chemical (phytohemagglutinin) that stimulates mitotic activity. After the cells multiply for one to two weeks, a mitotic spindle poison is usually added to the culture flask for several hours to arrest the dividing cells in metaphase. Without a mitotic spindle the chromosomes cannot proceed to anaphase, and thus cannot separate into daughter chromosomes. The cells are then washed in a low-salt (hypotonic) solution to make them swell and to disperse the bunched-up chromosomes. After fixation in acetic alcohol to preserve the chromosome structure, the cells are dropped onto microscope slides and stained. When a photograph is taken of a metaphase cell thus prepared, and the chromosomes are cut out of the print and lined up on the basis of size, we have what is called a karyotype (as in Fig. 6–1).

Suppose that a 40-year-old woman is newly pregnant and fears that her child may have Down's syndrome. Karyotyping of her fetus is now possible. Instead of taking a blood sample, cells sloughed off by the fetus into the amniotic fluid surrounding it in the uterus are used for the analysis. The process of withdrawing a sample of fetal cells from the amniotic fluid by means of a syringe passing through the abdominal wall of the pregnant woman is called amniocentesis (Fig. 6–9).

If amniocentesis is performed at 16 to 17 weeks after conception, there is enough time to culture the fetal cells in the laboratory, do a karyotype analysis, and find out if the fetus has Down's syndrome before the critical time for a therapeutic

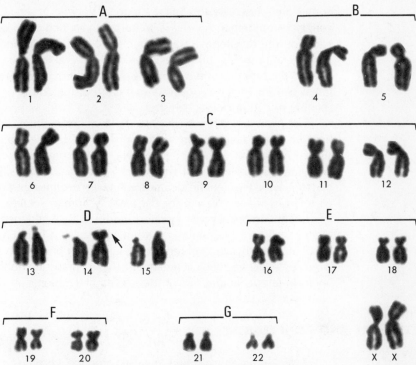

Figure 6–8 A karyotype of a 14/21 translocation. Note the extra arm on the righthand chromosome 14, indicated by the arrow. (Modified from Walker, Carr, Sergovich, Barr, and Soltan, J. Ment. Defic. Res. 7:150–163, 1963.)

abortion, if that is desired. Nowadays, a number of genetic defects, both chromosomal and chemical, can be detected before birth by tests on amniotic fluid samples, and many women do undergo an abortion if the test results are unfavorable for normal development (see Chap. Fifteen).

LETHAL GENETIC CONDITIONS

Many of the chromosomal abnormalities discussed here shorten the life span significantly. Over 90 per cent of Turner's cases die before birth. These fatal out-

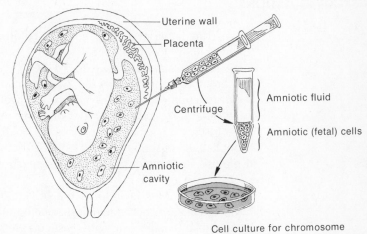

Figure 6–9 Amniocentesis for prenatal diagnosis of disease.

comes are caused by the lethal effects of genetic upsets. Ultimately, of course, all genotypes are lethal; no one is immortal. Inasmuch as all genes seem to have some effect on the development and maintenance of tissues and organs, every allele of every gene probably influences individual survival to some extent. Since no methods exist to measure the survival effects of this multitude of genes, in practice the term lethal genes refers only to those that can be singled out as causing death by some specific symptom or mechanism.

For example, albinism can cause death, but only after overexposure to sunlight or as the result of an accident triggered by the albino's typically poor eyesight. Death from albinism thus depends on environmental conditions, and indeed many so-called lethal genes are "conditional lethals." This is because the individual with the genetic defect may never run into an environmental factor that will be his or her undoing, or because, in human society, the environment of the individual can often be deliberately modified, as when medical treatment supresses a gene's effect. Hemophilia, an X-linked disease (Chap. Five), was unconditionally lethal at an early age until recently when a drug became available through research work for effective medical treatment. Now, many hemophiliacs survive to adulthood. In later chapters we will discuss other problems attendant on lethal genes.

SUMMARY AND CONCLUSIONS

Mistakes in mitosis and meiosis produce chromosomal abnormalities fairly often in humans and other organisms. The human phenotypes resulting from these changes in chromosome constitution often show severe malfunctions in several systems: muscle, skeleton, nervous tissue, and even reproduction may be impaired. Human sexuality may be particularly vulnerable to chromosomal upsets. Either the X and Y chromosomes are more commonly involved in aneuploid defects than the autosomes are, or sex chromosome aneuploidy just seems to be more common because its effects are less severe than the effects of autosomal aneuploidy. Recent research work has opened many doors to a finer understanding of the extent of chromosomal abnormalities throughout the life cycle and in different tissues. The next decade will undoubtedly witness an enormous enrichment of our ability to identify, predict, and minimize chromosomal defects.

STUDY QUESTIONS

1. Suppose the father of an XXY boy was colorblind and the boy was also found to be colorblind. How could this happen?
2. Some individuals are known to be XO/XY mosaics. What is the most likely sex chromosome makeup of the zygotes that gave rise to such individuals?
3. Nondisjunction can occur in meiosis II as well as meiosis I. How would the outcome differ for nondisjunction of one pair of chromosomes in meiosis I vs. II?
4. Some trisomy 21 females are fertile although the males are sterile (Table 6–2). About half the offspring of these females also have trisomy 21. What must happen in meiosis in terms of pairing and segregation of homologs to account for this 1:1 outcome in the F_1?
5. A few families are known to have a genetic defect which has a dominant effect in males but no effect in females. Affected males are XY, have testes *internally,* but look like females externally (testicular feminization syndrome). These males

are sterile; they lack an enzyme which permits the testosterone (male hormone) secreted by the testes in normal males to be converted to an active form. The active form permits male glands (prostate, etc.) to develop and function. Instead, in the affected males the reproductive system develops in the female direction. It is impossible to tell whether the causal testicular-feminization gene is autosomal or X-linked. Why?

6. Key terms:

sex determination	Down's syndrome	mosaics
Barr body	XO	translocation
Turner's syndrome	XXY	deletion
Klinefelter's syndrome	XYY	duplication
hermaphrodite	trisomy	transposition
nondisjunction	polyploidy	inversion
aneuploidy	triploidy	karyotyping
autosome	tetraploidy	amniocentesis
	lethal genes	conditional lethals

SUGGESTED READINGS

Clark, M. E. Contemporary Biology. Philadelphia: W. B. Saunders Co., 1973.
    Pages 593 to 608 are particularly good background for this chapter.
de Garay, H. L., L. Levine, and J. E. L. Carter (eds.). Genetic and Anthropological Studies of Olympic Athletes. New York: Academic Press, Inc., 1974.
    At the 1968 Olympics in Mexico City no chromosomal abnormalities were found in a sample of athletes, perhaps because the abnormalities are so rare, or because those with the abnormalities may have been eliminated prior to the games, or because "even slight differences from the chromosomal norm are incompatible with Olympic performance" (p. 224).
Kajii, T., K. Ohama, N. Niikawa, A. Ferrier, and S. Avirachan. Banding analysis of abnormal karyotypes in spontaneous abortion. American Journal of Human Genetics 25:539–547, 1973.
Masters, W. H., and V. E. Johnson. Human Sexual Response. Boston: Little, Brown, & Co., 1966.
    An objective and thorough description of human sexual activity.
Page, E. W., C. A. Villee, and D. B. Villee. Human Reproduction. Ed. 2. Philadelphia: W. B. Saunders Co., 1976.
Smith, D. W., and A. A. Wilson. The Child with Down's Syndrome. Philadelphia: W. B. Saunders Co., 1973.
    A paperback with much information for families with Down's children.
Stern, C. Principles of Human Genetics. Ed. 3. San Francisco: W. H. Freeman & Co., 1973.
    Gives extensive coverage of chromosomal abnormalities.
Swanson, H. D. Human Reproduction, Biology and Social Change. New York: Oxford University Press, 1974.

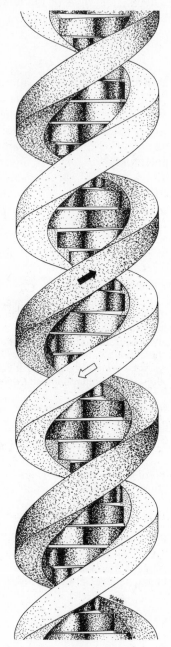

One visualization of the DNA molecule.

Chapter Seven

ATOMS TO ADAM

SOME CHEMICAL CONCEPTS

Some knowledge of chemistry can help illuminate how genes give rise to phenotypes. External phenotypes like albinism and "invisible" defects like colorblindness are both results of the chemical organization and functions of cells.

Most matter in the universe, from interstellar dust to cells, is made up of atoms, the smallest cohesive units of matter known. For simplicity, we may ignore the subatomic particles of the world of physics. Hydrogen gas in the galaxy and in the research laboratory contains two atoms of hydrogen (H) in each molecule of gas (H_2); the term molecule signifies a cohesive arrangement of atoms bonded together. Carbon dioxide molecules, which are produced by plants and animals and are found throughout the atmosphere, contain one atom of the chemical element carbon (C) and two of oxygen (O) for a total of three atoms (CO_2). Water molecules also have three atoms, two of hydrogen and one of oxygen (thus H_2O). Rocks, plastics, and especially living creatures have even larger numbers of atoms in some of their molecules.

Glucose, one of the simple sugars produced by plants, has 6 atoms of carbon, 12 of hydrogen, and 6 of oxygen, or $C_6H_{12}O_6$. Other sugars are also $C_6H_{12}O_6$, but they have the H's, O's, or C's in different positions from those found in glucose. For example, glucose differs structurally from fructose, the sugar found in many fruits:

Glucose   Fructose

The lines between the atoms represent the chemical bonds holding one C to another or to an O. The bonds attaching the hydroxyl groups (OH) or the hydrogens to carbons are not drawn, in order to accentuate the skeletal shape of the molecule.

Chemical bonds between atoms are of several kinds, all of which are important in holding various kinds of cell structures in a functional arrangement. Three kinds of these bonds are noteworthy: covalent bonds, ionic bonds, and hydrogen bonds. The bonds shown in the glucose molecule are covalent ones. In such bonds adjacent atoms in the molecule share subunits (electrons) of their atomic makeup. Carbon, oxygen, nitrogen, and many other kinds of atoms in cells are commonly held in place in molecules by covalent bonds.

Hydrogen is the smallest atom and is quite reactive with other atoms: sometimes it is bonded to other atoms covalently (as in water or glucose); sometimes it is shared in its entirety between two larger atoms (this is a hydrogen bond); and sometimes it acts as an electrically charged particle (in which case it is called an ion and written H^+). Atoms are electrically neutral: the ion H^+ is formed from a hydrogen atom (H) when it donates a subunit (electron) to a different atom. The recipient atom then becomes a negatively charged ion, for example, Cl^-, because electrons have a negative charge. $H^+ + Cl^-$, hydrogen ion plus chloride ion, form an ionic bond between them in the case of hydrochloric acid (HCl). Ionic bonds are formed between various small ions, such as H^+, OH^-, NH_3^+ (amino ion), and PO_4^+ (phosphate ion), as well as between the ionic parts of many kinds of large biological molecules.

The main chemical components of cells are water, inorganic elements like Fe (iron) and Mg (magnesium), carbohydrates (sugars and starches), lipids (fat-like substances), proteins, and nucleic acids (DNA and RNA, substances intimately involved with heredity). Carbohydrates, lipids, proteins, and nucleic acids are products of the living world. They are either made by each cell itself from smaller precursors or obtained as food, that is, as materials made in other cells. The ultimate source of all food for all cells on earth is certain simple materials of the environment that are fabricated into complex chemical structures inside green plants by the energy they capture from sunlight. By this process, called photosynthesis, plants produce sugars ($C_6H_{12}O_6$) and oxygen gas (O_2) from carbon dioxide gas (CO_2) and water (H_2O). By rearrangement of the chemical structure of sugar in subtle ways, starch and lipids, which contain only C, H, and O atoms, are synthesized in cells. By the rearrangement of sugar-like structures plus the addition of nitrogen (N), proteins are formed. They contain C, H, O, and N. By other rearrangements, plus the addition of both nitrogen and phosphorus (P), nucleic acids are formed. They contain C, H, O, N, and P. Proteins and nucleic acids are the largest molecules in cells. The chemical energy necessary to forge and maintain the chemical bonds in all these organic molecules inside cells comes ultimately from solar energy (through photosynthesis), from the "nuclear power" of the sun.

We must investigate the chemical relationships in cells to see how the process of gene → phenotype can be reduced to the process of DNA → RNA → protein, the "central dogma" of the biologist's understanding of genic activity. (The arrows in the dogma equation mean that one substance specifies the next, not that one turns into the next.)

PROTEINS

Proteins occur in all kinds of cells, many of which we find palatable and nutritious. All cells make their own proteins by linking together (polymerizing) small molecules called amino acids into various arrangements (many different proteins). Plant and animal cells manufacture, or synthesize, these amino acids essentially by adding nitrogen to breakdown products of sugar molecules. While plants are able to synthesize all the kinds of amino acids, some animals are not. Animals, including people, must obtain certain amino acids from their diet (the so-called "essential" amino acids).

An amino acid has the following general shape:

$$H_2N-\underset{\underset{R}{|}}{\overset{\overset{H}{|}}{C}}-COOH$$

where R symbolizes no one chemical element, but any one of many possible "side groups." The 20 amino acids we will be most interested in all have different R's and thus different overall structures. These 20 amino acids are the "links" in the chains we call proteins (Fig. 7–1). The size of proteins varies from about 6000 times the size of an H atom to over 1,000,000 times. That is, the molecular weights of proteins range from 6000 to over one million. To assemble these lengthy protein chains, covalent bonds are formed between individual amino acids at the cellular

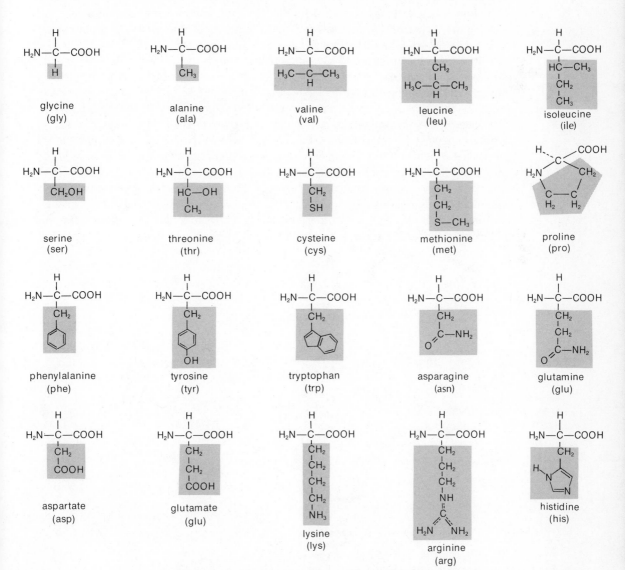

Figure 7–1 The twenty amino acids found in proteins.

factories called ribosomes. The synthesis of proteins is controlled by the genes (see Chap. Eight).

It is not known how many different proteins exist in an ordinary cell, but a variety of clues suggest there are at least 200 kinds. Each kind of protein is unique in its amino acid sequence. Because of these sequence distinctions, each kind of protein chain twists and curves around in space into a characteristic three-dimensional form (Fig. 7–2).

The functions of proteins fall into two main categories: structural and enzymatic. Structural proteins strengthen or protect other materials in the cell, chiefly in cell membranes, in the ribosomes, and in the chromosomes. Enzymatic proteins facilitate chemical reactions such as those which break down (degrade) foodstuffs, synthesize cellular parts, transport molecules through membranes, and perform other services rapidly and efficiently. There are well over 150 metabolic (degradative and synthetic) reactions in cells, and almost all are mediated, or catalyzed, by a specific enzyme. In the presence of an enzyme, a chemical reaction may proceed a million times faster than in its absence. Enzyme catalysis of chemical reactions is essential to permit the cell to grow and divide, to remove waste products, and so on. Life (as we know it) without enzymes is impossible.

It is not well understood how different enzymes bring about different chemical changes (for instance, converting glucose into fructose). Since each kind of protein has a different shape, however, each has a characteristic crevice or groove in its structure into which certain small molecules like glucose can fit. An active site is a groove on the enzyme where a molecule is broken down or is synthesized. Each kind of enzyme seems to have an active site which specifically fits its particular substrate, which may be a small molecule or a particular protuberance on a large molecule. Studies to date suggest that when a substrate molecule is lodged in the active site of the enzyme it becomes distorted. The resultant stress selectively breaks chemical bonds in the substrate, either cleaving it into smaller products or allowing the broken ends to react with other chemicals to create a new molecular form. In either case the enzyme ejects the products of the reaction once they are formed.

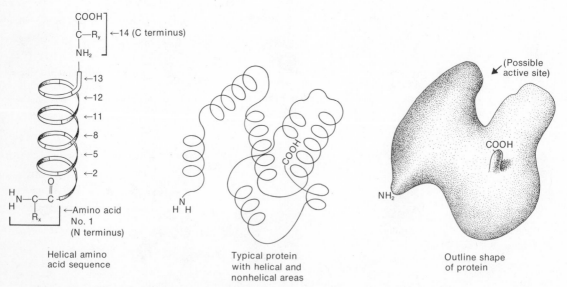

Helical amino
acid sequence

Typical protein
with helical and
nonhelical areas

Outline shape
of protein

Figure 7–2 Characteristic protein structure.

Enzymes are reusable a number of times, although like most everything in a cell they are ultimately degraded. Each kind of enzyme is probably represented in a cell by at least several copies. For some crucial enzymatic reactions (such as protein synthesis) many copies seem to be present at all times.

How enzymes find their appropriate substrates is not well understood. In general, enzymes probably come in contact with the appropriate substrate molecule by trial and error. Various small molecules in cells diffuse randomly throughout the cell interior. The higher their concentration, the more they will bump into their enzyme's active site and the faster the enzyme will act on them.

NUCLEIC ACIDS

Nucleic acids are the cellular molecules most intimately involved with heredity. The two kinds of nucleic acid are deoxyribonucleic acid (DNA) and ribonucleic acid (RNA). We now know that genes are composed of DNA, except in a few viruses where they are composed of RNA. Three kinds of cellular RNA, produced under the direction of DNA, function in the synthesis of proteins (Chap. Eight).

DNA molecules are absolutely the largest ones made by cells. Some are nearly a billion times the weight of a hydrogen atom. Each bacterial cell of the microorganism called *Escherichia coli* has only one kind of DNA molecule; that molecule is its entire chromosome. Human cells, by contrast, have 680 times the amount of DNA of an *E. coli* cell; toads and some other amphibians have even more DNA per cell than humans do (Table 7–1). How many different molecules of DNA each human cell contains is not known because the organization of plant and animal chromosomes at the level of molecules has not been determined, despite much effort. A cell with 46 chromosomes must have at least 46 DNA molecules, but it could have more if each chromosome contains many DNA molecules bonded together intricately.

All genes in human and bacterial cells are composed of DNA and all DNA molecules in all organisms are structurally alike in their general plan. They are long thread-like molecules built from the same three components: (1) phosphate (PO_4), which gives DNA its acidic properties; (2) deoxyribose, an unusual kind of sugar

TABLE 7–1   Estimates of Amount of DNA in a Haploid Set of Chromosomes in Various Organisms

Organism	Weight of DNA ($\times 10^{-12}$ grams)	Length of DNA in Terms of Nucleotide Pairs	Maximum Number of Genes*
Human	3.2	2,900,000,000	2,900,000
Cattle	2.8	2,500,000,000	2,500,000
Chicken	1.3	1,200,000,000	1,200,000
Toad	3.7	3,300,000,000	3,300,000
Carp (fish)	1.6	1,400,000,000	1,400,000
Snail	.7	600,000,000	600,000
Yeast (Saccharomyces)	.07	63,000,000	63,000
Escherichia coli	.005	4,200,000	4200
T_2 bacterial virus	.0002	180,000	180
øX174 bacterial virus	.000003	5500	5

*Assuming 1,000 nucleotide pairs per gene, that is, a protein product of about 330 amino acids in length.

with five carbons per molecule (ribose, the sugar in RNA, has an additional oxygen on one of its carbons as shown below):

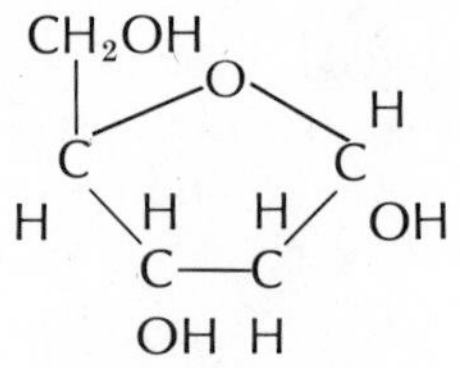

CH_2OH

Deoxyribose

CH_2OH

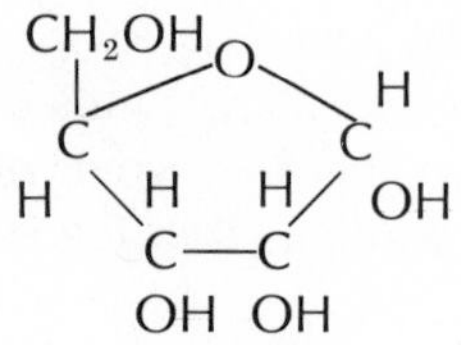

Ribose

and (3) nitrogenous bases, which give DNA some basic properties. (Household ammonia is a base and is reactive in a different way from, say, Clorox bleach, which is an acid.)

DNA has four kinds of nitrogenous bases. Two of the four have the double-ring purine structure, as shown below. These two are called adenine and guanine:

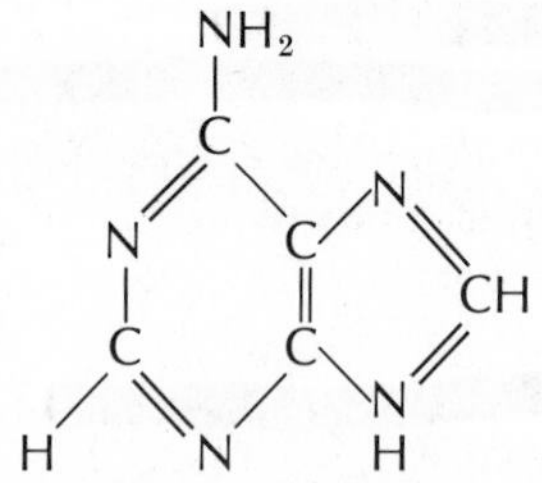

Memorize

Adenine (A)

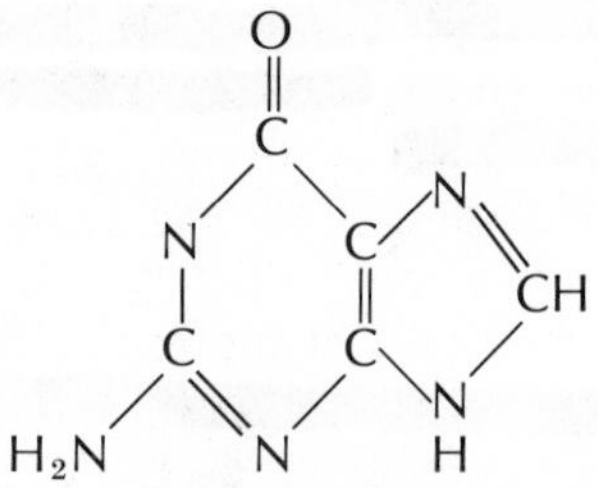

Guanine (G)

Adenine and guanine differ functionally because of their different side groups on top (NH_2 in adenine, O in guanine) and on the left side (H in adenine, and NH_2 in guanine), as shown. The purines are abbreviated A and G. The other two bases in DNA have the single-ring pyrimidine shape:

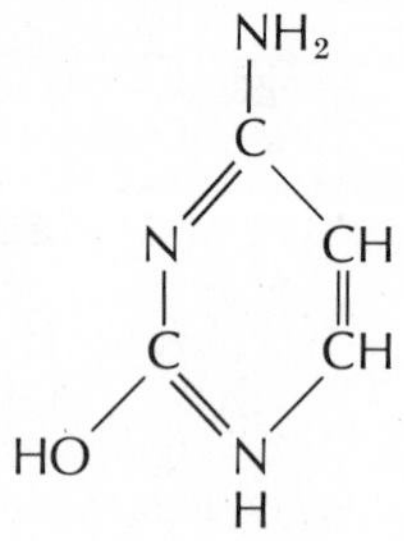

Cytosine (C)

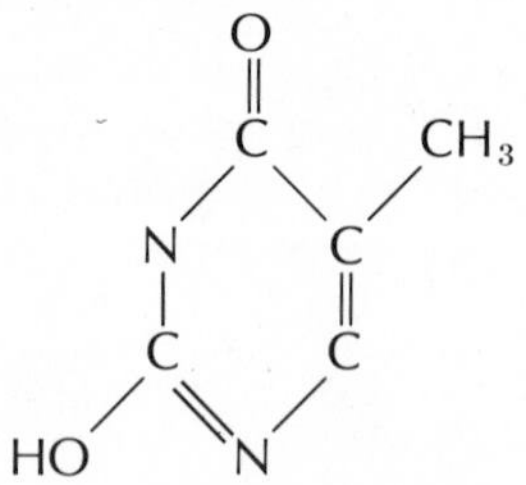

Thymine (T)

These two pyrimidines, cytosine and thymine, also differ functionally in DNA because of their different side groups. The pyrimidines are abbreviated C and T.

These components are put together as DNA subunits called nucleotides. Each nucleotide has the following composition: one phosphate attached to one sugar, which in turn is attached to one of the four types of bases:

PO$_4$
\
sugar
|
base   (A,T,G, or C)

A chain of these nucleotides looks somewhat like a charm bracelet, with the bases in seemingly random sequence, such as:

PO$_4$ ⟍sugar ⟋PO$_4$ ⟍sugar ⟋PO$_4$ ⟍sugar ⟋PO$_4$ ⟍sugar ⟋PO$_4$ ⟍sugar ⟋PO$_4$ ⟍sugar. . . etc.

| base | base | base | base | base | base |
| C | T | A | T | G | G |

The overall architecture of DNA was worked out in the 1950's only after it became suspected that the gene was DNA. Structural studies on DNA were prompted by biological discoveries of DNA function.

DISCOVERY OF GENE DNA

Although it had been known for a long time that chromosomes in cells contain DNA, it was not until the mid-1950's that the gene was conceded to *be* DNA. The reasons for this belated appreciation of DNA as the central molecule in biology are many, and they reveal much about how science works.

In the first place, chromosomes contain proteins as well as DNA. Since proteins are complexes of many (20) amino acids, many geneticists prior to the 1950's thought proteins were more likely candidates for gene stuff than DNA, which, most geneticsts thought then, was a small (not true), simple (not true) molecule composed of only four kinds of nucleotides (true). It was argued that if there was a gene "language," proteins with 20 "letters" in their alphabet had appealing versatility, compared to DNA with its four letter alphabet.

Secondly, the chemistry of DNA was poorly studied until about 1950—most biochemists had been spending their time unraveling the chemical form and function of carbohydrates, lipids, and proteins. The biological functions of these molecules were known before the functions of DNA and RNA were. Thirdly, the strong evidence developed from 1944 through the mid-1950's that DNA was the gene material was unknown to most geneticists of the day. Most geneticists were then concentrating on patterns of gene transmission in *Drosphilia,* corn plants, and other complex organisms. But the DNA studies were being done on simple microorganisms (bacteria and viruses) and were seldom published in the same scientific journals as studies done on corn.

One line of argument *in favor* of DNA being the genetic material did exist. Early in this century studies on cells had shown that DNA could be identified only in the chromosomes, whereas protein could be identified both in the chromosomes and throughout the cytoplasm of the cell. The idea of looking at DNA as the genetic material because it was the only kind of molecule uniquely found in chromosomes was appealing—but it was erroneous: we know now that DNA is also found in certain organelles in the cytoplasm (e.g., mitochondria), where it performs certain specialized hereditary functions. This raises the question whether mitochondria are descendants of bacteria-like cells that long ago took up a symbiotic residence inside a primitive mitotic cell (Chap. Four). For details of this topic, the reader may wish to refer to a longer text (e.g., Jenkins, see Chap. Two).

The landmark discovery that DNA had genic properties was published in 1944 by O. T. Avery and his co-workers. They showed that DNA isolated in the laboratory from cells of a strain of bacteria with trait "A" could enter cells of a strain lacking trait A and change, or transform, them into an A genotype (and phenotype). Avery showed that it was just the DNA in the donor cells, and not their RNA or protein, which had the power to transform the heredity of the recipient cells. This genetic transformation of the recipients was later shown to be due to the type-A DNA becoming part of the chromosome of the recipient cell (by an integrating process similar to crossing over, Chap. Five). The few reports of genetic transformation of animal cells by experimentally exposing them to foreign DNA have been viewed skeptically by many biologists. However, it is possible that certain viruses may transform mammalian cells into a cancerous state by the intregration of viral-type DNA into a mammalian chromosome (see Chap. Twelve).

Another important piece of information about DNA was developed in the later 1940's by E. Chargaff. By extracting DNA from many different organisms and analyzing its constituent bases chemically by an imaginative application of the technique called chromatography (Fig. 7–3), Chargaff showed that the DNA of organisms has the following common properties: (1) the number of purine residues in DNA approximately equals the number of pyrimidine residues, i.e., $A + G \simeq T + C$; (2) the number of A residues approximates the number of T residues ($A \simeq T$), and G approximates C ($G \simeq C$); and (3) the number of bases which have amino groups (NH_2) on top (as the structures were diagrammed previously) approximates the number which have oxygen (O) on top ($A + C \simeq G + T$). The significance of these findings, often called "Chargaff's rules," became clear only when J. D. Watson and F. R. C. Crick elucidated the structure of the DNA molecule. Avery's discovery of transformation in bacteria by DNA prompted Watson and Crick in 1953 to look critically at what little was then known about DNA structure and to try to understand how DNA could function as the central informational machinery for gene replication and gene control of cellular functions, as well as being the basis of gene mutations.

Watson and Crick gleaned current information about DNA structure from experiments done by physicists and chemists (see Box on Watson and Crick, p. 81), and after many setbacks succeeded in building a wire and metal model of the phos-

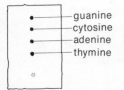

Figure 7–3 Chargaff's technique of separating the DNA bases of a degraded sample of DNA by paper chromatography.

1. The DNA sample is spotted onto a sheet of absorbent paper.

2. The lower edge of the paper is set in a dish of chemical solvent and held in place in an upright position.

3. The solvent diffuses up through the absorbent paper. When it is nearly to the top, the paper is removed and allowed to dry.

4. Special stains (or ultraviolet light) are applied to the paper to enable visualization of the location of the DNA. If the conditions have been well chosen, the DNA bases will be seen to have diffused up the paper at different rates (because of their chemical individuality) in the solvent. The bases will then be found in separate spots on the paper.

5. Next, the spots are cut out of the paper and each base redissolved and the amount present determined.

6. Results of experiment for DNA samples isolated from different organisms:

guanine
cytosine
adenine
thymine

	A	G	C	T
Human	.31	.19	.18	.31
Rat	.29	.21	.22	.28
Wheat	.27	.23	.23	.27
E. coli (bacteria)	.26	.25	.25	.24
Mycobacterium	.15	.35	.35	.15
T₂ bacterial virus	.33	.18	.17	.33

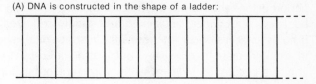

(A) DNA is constructed in the shape of a ladder:

(B) The DNA ladder contains 6 different subunits:

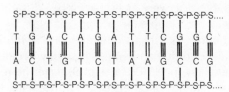

P, Phosphate
S, Deoxyribose sugar
A, Adenine (base)
T, Thymine (base)
G, Guanine (base)
C, Cytosine (base)
—, Covalent bond
=, Hydrogen bonds

Figure 7–4 The Watson-Crick model of DNA.

(C) The DNA ladder is twisted (helical):

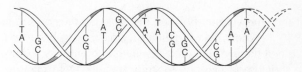

(D) The chain backbones run in opposite directions:

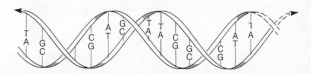

phates, sugars, and bases of DNA in correct three-dimensional relationship to each other. Their model had two nucleotide chains running in opposite directions with the bases of each chain on the interior and the two sugar-phosphate backbones on the exterior. A base on one chain is bonded to a base on the other chain by virtue of the fact that the two bases share hydrogen atoms between them (that is, the bases are paired by H bonds).

The overall shape of the Watson-Crick model of DNA resembles a ladder whose rungs are the so-called base pairs and whose side supports are the alternating phosphate-sugar backbones (Fig. 7–4). The ladder is, however, a twisted one, so that in three dimensions when upright it resembles a spiral staircase. The shape is technically called a double helix.

The helical shape, the sugar-phosphate side supports, the ladder rungs made of base pairs, and the kinds of bonds holding the structure together are common features of DNA in all organisms (just as the molecule glucose is the same shape in all organisms). But the DNA is different from one gene to the next in the same cell as well as being different from bacteria to pea plants. Its sole difference in all these contexts is in its sequence of base pairs for different genetic messages.

WATSON AND CRICK

In 1968 the American J. D. Watson wrote a personal account of the discovery of the structure of DNA *(The Double Helix).* The book describes the perceptions and social setting of Watson from 1951 through 1953, the period before and during the birth of the Watson-Crick model of DNA. The book was both widely acclaimed and widely criticized. On the one hand, it holds great historical interest as the first personal report by a participant in a major scientific discovery. But it also stirred up a hornet's nest of criticism because of Watson's sharp opinions of his colleagues and because of the dubious morality of his seizing onto the DNA project when others were already identified with it. Writing about the intellectual and personal drama of the DNA discovery as he remembered it more than a decade later, Watson states in *The Double Helix* that "many of the comments may seem one-sided and unfair, but this is often the case in the incomplete and hurried way in which human beings frequently decide to like or dislike a new idea or acquaintance" (Preface, p. 9). Later he refers to the problem of "jumping in on" studies of others by acknowledging that the scientific world is "complicated by the contradictory pulls of ambition and the sense of fair play" (Preface, p. 10).

At the time Watson finished his Ph.D. studies on the genetics of viruses, in 1950, he knew little biochemistry or physics because he had deliberately avoided them. Thereafter, his genetic mentor, S. Luria, and he agreed that he should go to Europe to study under a DNA chemist since the experiments by Avery in 1944 strongly suggested that DNA was the genetic material. Watson writes, "Of course there were scientists who thought the evidence favoring DNA was inconclusive and preferred to believe that genes were protein molecules. . . . Many were cantankerous fools who unfailingly backed the wrong horses" (Chap. 2, p. 18). But much as Watson dreamed about cracking the secret of the gene (DNA), he did not learn biochemistry during a whole year of intending to do so. Just as he was worrying about not accomplishing anything in Europe on gene chemistry, by good luck he heard a talk by the Englishman Maurice Wilkins on a confusing x-ray photograph of a crystal of DNA. The glimpse of DNA structure revealed by Wilkins suggested to Watson that the gene might have a straightforward, regular makeup and thus be a problem amenable to solution.

On the strength of this conviction, Watson migrated to Cambridge University in England with the intention of studying a "cross-cultural" discipline between physics and chemistry called x-ray crystallography. There he had the good fortune of meeting Francis Crick who, besides already knowing x-ray crystallography, turned out to be the effective stimulus and argumentative collaborator needed to accomplish the unveiling of DNA structure.

The story of the other participants in the DNA revolution—Erwin Chargaff, Linus Pauling, Rosalind Franklin, and, again, Maurice Wilkins—although important, can be skipped over here. The crux of the problem was solved by Watson and Crick. Homing in on Pauling's recent demonstration of helical structure in proteins, and on the Wilkins-Franklin suggestion of exterior backbones and two or three helical chains in DNA, Watson and Crick were nevertheless stumped for months on how the chains fit together. Finally, after many false solutions, Watson hit on the exquisitely beautiful idea of a fit between particular bases in hydrogen-bonded base pairs: the idea immediately revealed how the two chains were combined and also how gene replication could occur with precision. Together, Watson and Crick conceived of these implications as they built a precise metal model of a segment of DNA with its complementary base pairing of A:T, T:A, G:C, and C:G, an arrangement validating Chargaff's rules about the relative proportions of the bases in DNA isolated from many organisms.

James Watson and Francis Crick, along with Maurice Wilkins, received the Nobel prize in 1962 for their work on DNA. (See Box on Nobel Prize, below).

Let's back up a minute and look more closely at DNA bases and base pairs. Recall that among the four DNA bases (A,T,C, and G), the purines A and G are similar in shape and differ only in certain small side groups; likewise the pyrimidines T and C are also similar, differing again only in side groups. These variable side groups are crucial to understanding base pairing in DNA. The fact is that the side groups of A are complementary *only* to those of T when both are "welded" into the DNA double helix on opposite strands. The same goes for G and C: they can also bond in complementary, dovetailing fashion between the two strands in DNA. That is, if G is a "jigsaw" piece on one strand of DNA, it can only pair up with the complementary piece called C on the opposite strand. Likewise, C on one strand only interdigitates (or pairs) with G on the other strand. The four kinds of base pairs between the two helical backbones in DNA are thus A:T, T:A, G:C, and C:G. If an unaffiliated single strand of DNA existed that had the base sequence

$$\overrightarrow{T\ G\ C\ C\ T\ A\ A\ G\ C}$$

it could pair up correctly with any available single strand of the complementary base sequence

$$\underleftarrow{A\ C\ G\ G\ A\ T\ T\ C\ G}$$

but with no other sequence.

Overwhelming experimental evidence suggests that the sequence of these four base pairs in a DNA molecule is the chemical code, or message, for all the inherited information in cells. How this simple coding mechanism functions in determining the structures and activities of cells will be described in the next chapter.

THE NOBEL PRIZE

Many research workers in genetics have received the Nobel prize, particularly those who have worked at the level of DNA, RNA, and proteins. Among these Nobel laureates are the following, cited by year of award:

1910   A. Kossel for studies on the chemical composition of the cell nucleus.

1930   K. Landsteiner for work on genetic differences in blood types (see Chap. Eleven).

1933   T. H. Morgan for work on chromosomal mechanisms of gene transcription.

1946   H. J. Muller for discovery of the induction of mutations by x-rays (see Chap. Thirteen).

1958   G. Beadle and E. Tatum for work on gene activity in metabolism

(Chap. Ten); and J. Lederberg for work on genetic recombination in bacteria.

1959   A. Kornberg and S. Ochoa for the artificial synthesis of DNA and RNA, (Chap. Eight).

1962   J. Watson, F. Crick, and M. Wilkins for work on the nature of the DNA molecule (Chap. Seven).

1965   F. Jacob, J. Monod, and A. Lwoff for studies in the genetics of bacteria (Chap. Nine) and viruses (Chap. Twelve).

1968   M. W. Nirenberg and H. G. Khorana for deciphering the genetic code; and R. W. Holley for elucidating the structure of transfer RNA (Chap. Eight).

1969   M. Delbruck, S. Luria, and A. D. Hershey for pioneering work on the genetics of bacterial viruses.

1972   R. Porter and G. M. Edelman for work on the structure of antibodies (Chap. Eleven); and C. B. Anfinson, S. Moore, and W. H. Stein for studies on the structure and activity of the enzyme ribonuclease.

1975   R. Dulbecco, H. M. Temin, and D. Baltimore for studies on the interaction of tumor viruses and cellular DNA (Chap. Twelve).

The Nobel prize is named after Alfred Nobel, the Swedish inventor of dynamite. He was a pensive man and sensitive to criticism that dynamite was a destructive force in the world. (Although now that we have atomic bombs, dynamite sounds like child's play.) Having amassed great wealth during a lifetime of scientific inventiveness and business success, Nobel set up the prizes in his will to recognize and encourage humanitarianism.

The Nobel prizes were first awarded in 1900. They are given in the categories of literature, peace, economics, and several natural sciences. Most geneticists who are Nobel laureates have won the prize in "physiology or medicine," there being no category labeled genetics. The natural sciences were included among the prizes by Nobel because he believed that science is a constructive enterprise for bettering the human condition. But the "courtly and secretive" Swedish professors who decide on Nobel prize winners in the sciences have been criticized for ignoring fields like mathematics, agriculture, and psychology.

The Nobel prizes are the best known awards for scientific achievement, and among the most lucrative. Much has been said and written about the anxiety, excitement, rivalry, and envy among scientists concerning the Nobel awards; it is the subject of a novel written by Irving Wallace (*The Prize*: Simon and Schuster, 1962).

SUMMARY AND PERSPECTIVE

Cells, and therefore multicellular organisms such as humans, are chemical machines. Like all other chemical structures in nature, cells obey the rules of chemistry: they are composed of atoms bonded together to form molecules of various sorts that have specific cellular functions. These cellular molecules can be isolated and examined in the laboratory to study their individual structures and their interactions with other molecules. But living cells possess extraordinary properties not exhibited by mixtures of molecules in the laboratory. Molecules in cells are organized into supra-molecular relationships (membranes, chromosomes, etc.) which confer on cells special properties (growth, repair, reproduction) that permit the phenomena collectively called life to occur.

Life is chemically a unity—all cells being composed of the same materials—although in outward form and behavior life is very diverse. Both the likenesses among cells and the diversity among organisms are basically the product of the chemical substance DNA. DNA is the hereditary material in cells that has a marvelously simple chemical code embedded in its structures that determines both the "skin in" chemical unity and the "skin out" diversity of life.

STUDY QUESTIONS

1. What is the source of materials for cellular structure in animals? Plants?
2. What is the source of energy for building and maintaining cell structure? For making and breaking chemical bonds in cell constituents?
3. Where are proteins found in cells? What are the functions of proteins?
4. What is wrong with the DNA in the "straight-ladder" diagrams below?

(a) $\overrightarrow{\text{T C} \quad \text{C A} \quad \text{G T} \quad \text{C A}}$ (b) $\overrightarrow{\text{T C A C A G G T T C A}}$

$\underleftarrow{\text{A G} \quad \text{G T} \quad \text{C A} \quad \text{G T}}$ $\underleftarrow{\text{A G T G T C C A G G T}}$

5. Key Terms:

atom	photosynthesis	adenine (A)
molecule	amino acids	guanine (G)
chemical bond	metabolic	cytosine (C)
covalent bond	enzyme	thymine (T)
ions	active site	nucleotide chain
ionic bond	DNA	transformation
hydrogen bond	nitrogenous bases	Watson-Crick model
carbohydrates	purine	double helix
lipids	pyrimidine	base pairs
proteins		

SUGGESTED READINGS

Evlanoff, M., and M. Fluor. Alfred Nobel, The Loneliest Millionaire. Los Angeles: Ward Jenkins, Ritchie Press, 1969.

Koshland, D. E., Jr. Protein shape and biological control. Scientific American 229:52–64, 1973.

Moore, S. and W. H. Stein. Chemical structure of pancreatic ribonuclease and deoxyribonuclease. Science 180:458–464, 1973.

Nobel Foundation and W. Odelberg (ed.). Nobel, the Man and His Prizes. New York: American Elsevier Publishing Co., Inc., 1962.

Olby, R. The Path to the Double Helix. Seattle: University of Washington Press, 1975.

Pauling, P. The race that never was? New Scientist 58:558–560, 1973.

Sayre, A. Rosalind Franklin and DNA. New York: W. W. Norton & Co., Inc., 1975.
    A story which illustrates the problems of women in science.

Stent, G. S. Prematurity and uniqueness in scientific discovery. Scientific American 227: 84–93, 1972.
    A discussion of Avery, Watson, and Crick.

Stroud, R. M. A family of protein cutting proteins. Scientific American 231:74–88, 1974.

Watson, J. D. The Double Helix. New York: Atheneum Publishers, 1968.

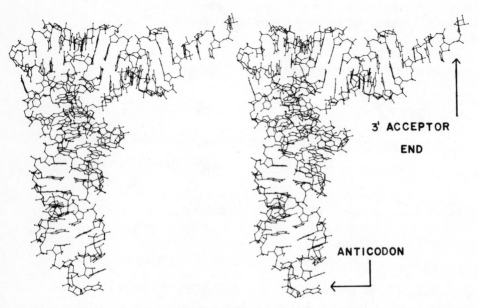

Stereoscopic drawing of transfer RNA. Look directly at figure, letting your eyes relax, in order to see the three-dimensional structure. (From Quigley, G. J., and Rich, A.: Structural domains of transfer RNA molecules. Science, *194*: 796, 1976, p. 798.)

Chapter Eight

GENE ACTIVITY

The idea that DNA is the gene material, brilliantly elucidated by Watson and Crick, was quickly and effectively upheld in a multitude of studies after 1953. The dogma that DNA is both a self-replicating molecule and a purveyor of chemical information for cell functions was soon reduced to the formula DNA $\rightleftharpoons$ DNA $\rightarrow$ RNA $\rightarrow$ protein. We will consider the segment DNA $\rightleftharpoons$ DNA first, that is, the process of DNA replication and, therefore, gene replication.

DNA REPLICATION (DNA $\rightleftharpoons$ DNA)

As Watson and Crick predicted, and as A. Kornberg, M. Meselson, F. Stahl, and others showed, DNA replicates by a process of double copying of the two strands of the DNA molecule. Since the original two strands are complementary to each other in their base sequence, the two daughter strands are complementary to the template strands they are copied from, and thus also complementary to each other (Fig. 8–1).

The two strands of a daughter DNA molecule are not both new—one of them is new (the copy) and one is old (the template). In replication, DNA molecules are thus "semi'conserved"; as the two original strands unwind and each is copied, a new strand curls around each old strand. As this implies, the base pair sequence of both daughter molecules is the same as the parental molecule, assuming that no errors have occurred in the copying process. (Such errors in replication cause DNA base mutations, discussed in Chap. Thirteen.)

According to recent studies by Arthur Kornberg and others, a whole complex of enzymes is necessary to accomplish DNA replication. To date, at least seven enzymes have been identified which take part in the steps of replication. These steps include (1) initiation of synthesis of new DNA strands, (2) strand elongation, and (3) separation. Initiation involves the binding of a certain "unwinding protein" to the old DNA along its entire length except at the places where new strands will start to grow (the top ends in Fig. 8–1). Under the influence of an RNA-promoting enzyme, a small piece of RNA then attaches to the exposed initiation spot (this role of RNA is a recent finding and is still under investigation). The RNA acts as a sort of "sky hook" to which DNA precursors then attach to start building the new strands. Strand elongation proceeds as precursor pieces align themselves correctly on the templates of the old strands by proper base pairing through hydrogen bonds (A opposite T, C opposite G, and so on). A DNA polymerase enzyme bonds together the precursor pieces of the new strands as the old strands "unzip" and new precursors

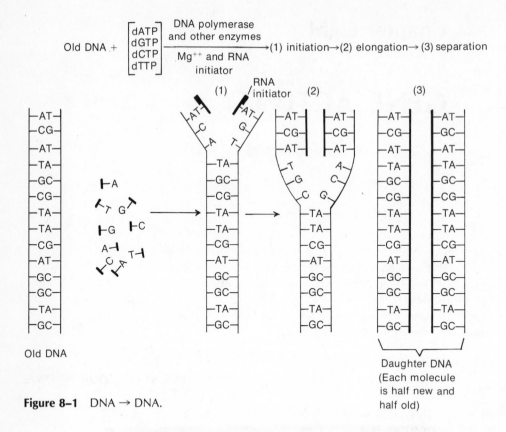

Old DNA $+$ $\begin{bmatrix} dATP \\ dGTP \\ dCTP \\ dTTP \end{bmatrix}$ $\xrightarrow[\text{Mg}^{++}\text{ and RNA initiator}]{\text{DNA polymerase and other enzymes}}$ (1) initiation→(2) elongation→ (3) separation

Figure 8–1   DNA → DNA.

Daughter DNA
(Each molecule
is half new and
half old)

are laid down. When the new strands are finished, the two daughter molecules separate completely. The genetic content is now doubled.

The replication of DNA in large chromosomes (larger than those possessed by viruses) probably involves many initiation sites per chromosome. The process described above would then produce daughter DNA molecules that have gaps in the new strands because the DNA polymerase enzymes can only add on precursors in one direction. When a polymerase enzyme has worked its way down to another initiation site which already has a fragment growing out from it, the enzyme cannot "bridge the gap" to the next piece (Fig. 8–2). However, another enzyme, called ligase, has the ability to seal up single-strand gaps between fragments (which are thought to be about 1000 nucleotides long, according to the work of Okazaki). Thus a continuous strand is produced by the combined action of polymerase and ligase enzymes.

DNA that is biologically active can be synthesized in a cell-free system (a mixture of ions, enzymes, precursors, and a piece of DNA "primer" or template).

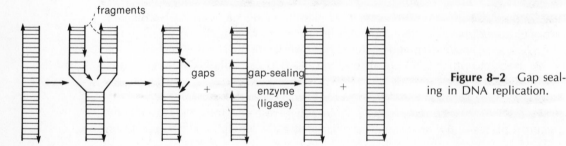

Figure 8–2   Gap sealing in DNA replication.

RNA TRANSCRIPTION (DNA → RNA)

Although RNA has been mentioned in passing, we have waited until now to give it its proper attention because its biological significance follows from the fact that it is synthesized on a DNA template.

RNA is usually a single-stranded nucleic acid molecule. It is composed of the same subunits as DNA, except that RNA contains ribose sugar instead of deoxyribose and the base uracil rather than thymine (Table 8–1).

RNA molecules under the majority of circumstances are synthesized as copies of DNA; this process is called transcription, while the term replication is reserved for DNA copying DNA. Many different RNA molecules are transcribed from one DNA molecule, each molecule of RNA representing one (or a few) genes. RNA transcription occurs by the same base-pairing rules that are operative when new strands of DNA are synthesized. The three main differences between DNA replication and RNA transcription are: (1) an adenine (A) on a DNA template strand specifies a uracil (U) on the RNA copy strand; (2) RNA is copied from only one of the two DNA strands; and (3) enzymes mediating transcription (RNA polymerases) differ from those that mediate replication (DNA polymerases).

TABLE 8–1   Comparison Between RNA and DNA

	RNA	DNA

Characteristics	Single stranded (usually)
	Found in all cells
	    and some viruses
		Double stranded (usually)
		Found in all cells
		    and other viruses

Components	Phosphate Ribose sugar: Adenine Guanine Cytosine Uracil:	Phosphate Deoxyribose sugar: Adenine Guanine Cytosine Thymine:
Location in cells	Nucleus Cytoplasm	Nucleus Cytoplasmic DNA scanty
Functions	Genic (in viruses) Messenger RNA Transfer RNA Ribosomal RNA Special functions	Genic (in viruses and cells) Special functions (possibly)

Transcription: DNA→RNA

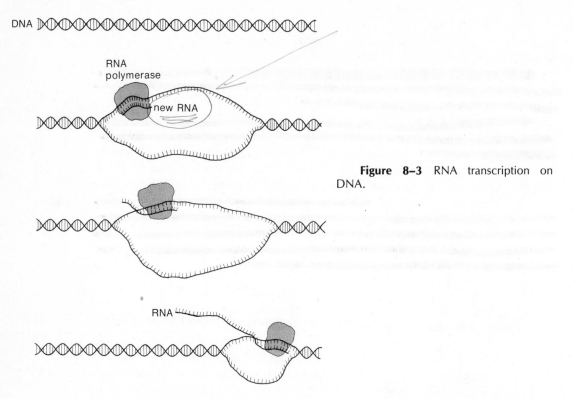

Figure 8–3 RNA transcription on DNA.

Since RNA copies of DNA are shorter than the whole DNA molecule, we can envision the process as one in which the DNA "zipper" gaps open for a stretch to permit RNA synthesis there (Fig. 8–3).

If the DNA strand being copied via the RNA polymerase reads

T G C C T A A G C
_____

the RNA produced on this template surface will read

A C G G A U U C G
_____

as a result of Watson-Crick base pairing.

Although the activity of RNA molecules is a complex subject, it is sufficient here to point out the three major kinds of RNA and their functions. The three kinds are messenger RNA (mRNA), transfer RNA (tRNA), and ribosomal RNA (rRNA). All of them, after being synthesized on a gene (DNA), peel off their template and move away (ultimately to the ribosomes) to perform their specialized functions.

Messenger RNA is a complementary copy of the DNA base sequence in ordinary genes, that is, ones that determine the structure of specific proteins such as enzymes. The mRNA product of such a gene moves to the cytoplasm where it attaches to a ribosome and directs the polymerization of amino acids in the proper order to form a particular protein. After being reused a number of times in the ribosome "factories," each time producing the same protein product, the mRNA molecule is degraded by normal cellular processes. Genes can and often do produce

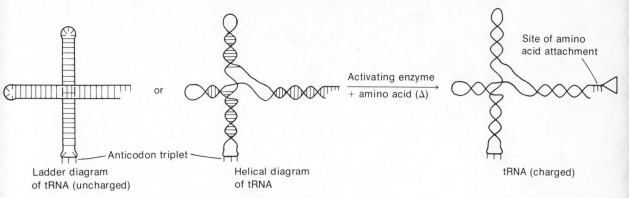

Figure 8–4   Transfer RNA activation.

multiple copies of mRNA transcripts for brisk synthesis of their protein product on ribosomes. A cell has the ability to synthesize as many kinds of mRNA transcripts as there are genes that determine different proteins.

Transfer RNA comes in only 20 or so varieties. There is at least one kind for each of the 20 amino acids found in proteins. tRNA molecules are small, single-stranded copies of special tRNA genes; tRNA gene products are never used as an RNA message for translation into the amino acid sequence of a protein. One reason for this lack of translation of tRNA's may be their complex shape. Whereas mRNA molecules are uncoiled, tRNA's are coiled back on themselves in a cloverleaf shape (Fig. 8–4). One of the free ends on a tRNA molecule is the site where an amino acid can be attached with the aid of a specific enzyme (activating enzyme). Once the amino acid (aa) is attached, the tRNA is said to be charged or activated. On the ribosome, activated tRNA's (tRNA–aa) transfer their dependent amino acid to a growing chain of amino acids which, when finished, will constitute a protein. The control of which amino acid will be in which spot on the protein chain is determined by the matching of another site on tRNA molecules, called the anticodon, with the mRNA on the ribosome. The anticodon is the end of one of the "switch back" loops of the tRNA structure and consists of three unpaired bases (of various sequences). More about this later.

Ribosomal RNA is of two kinds, each being the product of a specific gene in cells. These rRNA-producing genes, like those genes that produce the diverse tRNA's, are also unusual in that their products are not translated into protein structure. rRNA molecules are part of the makeup of ribosomes and help form the ribosomal housing in which protein synthesis occurs. The rest of the ribosome consists of a number of proteins (ribosomal proteins).

PROTEIN SYNTHESIS (RNA → PROTEIN)

Genes make more genes (DNA), as well as determining the formation of non-genes: RNA and proteins. Both rRNA and tRNA are end products of gene activity—their base sequence does not contain information that is translated into proteins, although they have indispensable functions in protein synthesis. mRNA, on the other hand, is translated into the amino acid sequence of proteins. This occurs in a devious fashion.

We require at least a picture (Fig. 8–5) to see how nucleotides are translated

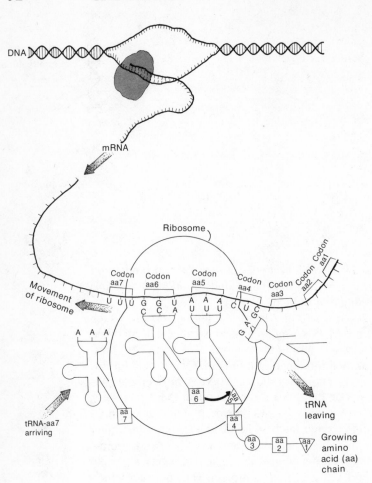

DNA

mRNA

Ribosome

Figure 8-5   Protein synthesis.

Codon
aa7

Codon
aa6

Codon
aa5

Codon
aa4

Codon
aa3

Codon
aa2

Codon
aa1

Movement
of ribosome

U U U   G G U   A A A   C   U C
         C C A   U U U   G   U
                               G G
A   A   A

aa
6

aa
7

aa
5

aa
4

tRNA-aa7
arriving

tRNA
leaving

aa
3

aa
2

aa
1

Growing
amino
acid (aa)
chain

into amino acids, that is, how proteins are synthesized. A motion picture would be even better to visualize the process, which is dynamic and of multiple moving parts. The energy for this process is provided by ATP and related molecules (not shown in Fig. 8–5). Various kinds of proteins are needed to initiate the synthesis of new proteins, to elongate the growing chains of amino acids, and to permit the release of the finished chains from the ribosomes. Our knowledge of this process is the result of many laboratory experiments over a long period (since 1961), utilizing cell-free (in vitro) systems, that is, those which combine isolated ribosomes, mRNA, tRNA, particular enzymes, and other ingredients in a test tube situation.

Figure 8–5 indicates how protein molecules are synthesized: the mRNA transcript of a gene moves away from the chromosome and attaches to a ribosome along its "waist." The tRNA molecules charged with appropriate amino acids attach to the ribosomes at specific sites, two tRNA's filling the sites on a ribosome at any one time. A tRNA–amino acid complex can only attach to a ribosome if the anticodon—the exposed three bases on one end of the tRNA—base pairs correctly with the codon, the three base sequence on the mRNA. For example, an anticodon of sequence AUA could hydrogen bond with the codon triplet UAU on the ribosome-bound mRNA. This pairing between mRNA and tRNA positions the amino acid attached to the tRNA "tail" correctly on the ribosome surface. The amino acid can then be joined to another amino acid to form a nascent protein chain. As one amino acid bonds to the next, its carrier tRNA is released from the ribosome and the ribosome trundles along the mRNA base sequence so as to expose the next codon in line to attack by a tRNA anticodon. tRNA's thus move in on one side of the ribosome, "read" the mRNA codon, shift to the chain-holding site, and then move off on the other side of the ribosome. By repeated cycling of tRNA's onto and off of the ribosome, the amino acid chain grows and grows. When a ribosome reaches the end of the mRNA, it falls off. The finished protein is then released. It folds up into its characteristic shape and begins its functional existence as an enzyme, membrane component, or other integral part of the cell.

Protein synthesis is the fundamental process for gene guidance of cellular activities.

THE GENETIC CODE

The relationship between particular base sequences in DNA and RNA and particular amino acid sequences in proteins is determined by the genetic code (Fig. 8-6). The genetic code was deciphered by a variety of in vitro studies on protein synthesis utilizing molecular components from microorganisms. (See Box on Deciphering the Genetic Code, p. 94.)

The base sequence in the DNA of the gene, and therefore its mRNA transcript, are co-linear with the amino acid sequence in the protein product. Suppose we had the following amino acid sequence (reading from the amino end):

valine–histidine–leucine–threonine–proline–glutamic acid. . . . . . .

It could be coded by an mRNA sequence of the following codons:

G U A · C A U · C U C · A C U · C C C · G A A. . . . . .

(or of various other codons with different letters in the third position). Such an mRNA sequence would be produced by DNA of the following base sequence:

C A T G T A G A G T G A G G G C T T. . . . . .
G T A C A T C T C A C T C C C G A A. . . . . .

Figure 8–6 The genetic (mRNA) code for proteins. The codons at issue do not translate into any amino acid in protein; they signal the termination of protein synthesis.

UUU ⎤ phe	UCU ⎤	UAU ⎤ tyr	UGU ⎤ cys
UUC ⎦	UCC ⎪ ser	UAC ⎦	UGC ⎦
UUA ⎤ leu	UCA ⎪	UAA*	UGA*
UUG ⎦	UCG ⎦	UAG*	UGG try
CUU ⎤	CCU ⎤	CAU ⎤ his	CGU ⎤
CUC ⎪ leu	CCC ⎪ pro	CAC ⎦	CGC ⎪ arg
CUA ⎪	CCA ⎪	CAA ⎤ gln	CGA ⎪
CUG ⎦	CCG ⎦	CAG ⎦	CGG ⎦
AUU ⎤	ACU ⎤	AAU ⎤ asn	AGU ⎤ ser
AUC ⎪ ile	ACC ⎪ thr	AAC ⎦	AGC ⎦
AUA ⎦	ACA ⎪	AAA ⎤ lys	AGA ⎤ arg
AUG met	ACG ⎦	AAG ⎦	AGG ⎦
GUU ⎤	GCU ⎤	GAU ⎤ asp	GGU ⎤
GUC ⎪ val	GCC ⎪ ala	GAC ⎦	GGC ⎪ gly
GUA ⎪	GCA ⎪	GAA ⎤ glu	GGA ⎪
GUG ⎦	GCG ⎦	GAG ⎦	GGG ⎦

This particular DNA sequence may exist among human genes because: (1) the genetic code is known to be a universal feature of DNA (occurring in all organisms); and (2) the amino acid sequence given above belongs to the first part of the gene product for the beta chain of hemoglobin (the oxygen-carrying protein in red blood cells).

DECIPHERING THE GENETIC CODE

Step 1. Obtaining the correct cellular ingredients for in vitro protein synthesis: the components isolated from microorganisms that proved to be useful included ribosomes, a full complement of tRNA's, and the enzymes for attaching amino acids to them.

Step 2. Obtaining an mRNA with known base sequence: laborious chemical studies by G. Khorana and others yielded the following purified artificial mRNA's (and many more):

Poly U   (U–U–U–U–U–etc.)   poly C   (C–C–C–C–C–etc.)

poly A   (A–A–A–A–A–etc.)   poly U with single C (C–U–U–U–U–etc.)

Step 3. M. Nirenberg and co-workers first showed that a test tube combination of the ingredients in Step 1 and certain shelf chemicals (ATP, amino acids, and magnesium) with one of the known-sequence mRNA's in Step 2 gave informative results. For example,

(Exp. 1)

$$\begin{bmatrix} \text{ribosomes} \\ \text{tRNA's} \\ \text{enzymes} \end{bmatrix} + \begin{bmatrix} \text{amino acids} \\ \text{ATP} \\ \text{magnesium} \end{bmatrix} + \text{poly U} \rightarrow \text{protein consisting only of phenylalanine}$$

  (from cells)   (stock chemicals)

 Interpretation: UUU in mRNA codes for phenylalanine in protein.

(Exp. 2)

[as above] + [as above] + poly C → protein consisting only of proline

 Interpretation: CCC in mRNA codes for proline in protein.

(Exp. 3)

[as above] + [as above] + poly U with end C → protein consisting of leucine followed by many phenylalanine residues

 Interpretation: CUU in mRNA codes for leucine in protein.

Genetic variants of normal hemoglobin (Hb A) are known that differ from Hb A in only one amino acid in the sequence. For example, sickle-cell hemoglobin (Hb S), which we will discuss further in Chapter Ten, has valine instead of glutamic acid in the sixth position from the amino end. The information in the genetic code makes clear that such a genetic change could have occurred by a mutation in a single base pair in an ancestor's DNA, since one mRNA codon for valine is GUA (Fig. 8–6) and one for glutamic acid is GAA. The DNA for Hb A at this site would then read $\frac{CTT}{GAA}$ and for Hb S would read $\frac{CAT}{GTA}$. The phenotypic consequences of this minute change in DNA (and hemoglobin) are disastrous; sickle-cell homozygotes usually die of anemia at an early age.

Research workers have now identified within the human population more than 100 different amino acid substitutions in the protein hemoglobin. By studying the physical shape and chemical reactions of these genetic variants of hemoglobin, scientists are learning more about the ways in which the molecule functions in the blood. Application of this knowledge for devising medical treatment for sickle-cell anemics, and others with hemoglobin defects, is currently moving forward at great speed.

The genetic code was worked out by patient and dedicated study in many laboratories during the 1960's. It holds for gene translation into proteins in all kinds of organisms studied (Fig. 8–7). It explains why many genetic variants of proteins differ from the normal protein in a single amino acid, since the mRNA code is "read" by tRNA's in linear fashion, three bases at a time. Laboratory studies have shown that in many cases the change of just one base specifies the inclusion of a different amino acid in the protein chain. Note, however, that the change from AAA to AAG at the DNA level (UUU to UUC for mRNA) does not change the protein product (Fig. 8–6). In many instances the genetic code "buffers" the protein product against excessive change because the code possesses redundancy. For example, in some cases (such as for leucine) as many as six different codons specify a single amino acid. Should a mutation occur in the third position of the mRNA CU– codon for leucine, no harm will befall the decoded protein; that is, it will be a normal protein, not a mutant one.

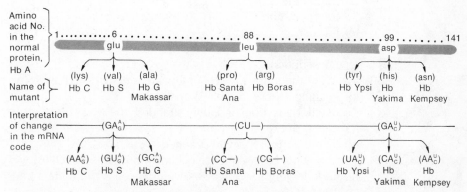

Figure 8–7 Demonstration of agreement between some of the known human hemoglobin variants (of the so-called beta chain) and one-step substitutions in mRNA codons. Such agreements demonstrate one applicability of the genetic code to human genes. (Hemoglobin variants are often named for the place where they were first found.)

TABLE 8–2 Summary of Gene Activity

	Process	Location
Replication	DNA → DNA	Nucleus
Transcription	DNA → RNA	Nucleus
Translation	RNA → protein	Ribosomes

SUMMARY

The DNA story is a "cathedral" in biology. In the past two decades it has revolutionized the way we think about how life functions. The microscopic molecules of DNA are self-perpetuating chemical devices capable of replication and also of producing RNA copies: a few of the copies are ribosomal-specific and transfer-specific RNA's, but most RNA molecules are messenger RNA's. Messenger RNA directs the synthesis of protein on the numerous subcellular ribosomes. Proteins, the final products of most genes, are the immediate functionaries in all cellular activities through their enzymatic and structural properties (Table 8–2).

DNA control of cellular activities occurs through the delicate matching of mRNA base sequences with areas on transfer RNA molecules. This process specifies particular amino acid sequences in proteins. Transfer RNA molecules are the versatile "double-handed employees" for protein translation from the genes: a tRNA recognizes an individual amino acid as well as an individual triplet codon of mRNA.

The genetic code is the mechanism by which particular nucleic acid base sequences are translated into particular amino acids to make proteins. Nucleic acid code words are triplet base sequences read in a linear fashion for protein synthesis in the same way in all organisms.

STUDY QUESTIONS

1. Discuss whether DNA, like mRNA (as discussed in the text), is degraded "after being reused a number of times." Do you think enzymes are degraded after being used a number of times? Why or why not?
2. The synthesis of RNA on a DNA template is known as
   (a) translation
   (b) transportation
   (c) transfer
   (d) transformation
   (e) transcription
3. Which of the following sets of terms correctly describes RNA?
   (a) guanine, ribose, double stranded
   (b) uracil, deoxyribose, double stranded
   (c) uracil, ribose, single stranded
   (d) thymine, ribose, single stranded
4. The following base sequence occurs in one strand of DNA: <u>ATTGACCTC</u>. Which of the following suggested RNA sequences is the correct transcription product?
   (a) TAACTGGAG
   (b) AUUGACCUG
   (c) GAGGTCAAT
   (d) UAACUGGAG
   (e) CGGUCAAGA

5. Assuming all base pairs code for amino acids, a DNA segment (gene) 3000 base pairs long would, through transcription and translation, determine a protein that is how many amino acids long?
   (a) 3000
   (b) 9000
   (c) 1000
   (d) 100
   (e) 300

6. The anticodon refers to a region in which of the following?
   (a) DNA
   (b) transfer RNA
   (c) messenger RNA
   (d) ribosomal RNA
   (e) ribosomes

7. Supposing an mRNA molecule is produced with the sequence CACACACACACA . . . . etc. How many different kinds of amino acids will the protein product contain?
   (a) 0
   (b) 1
   (c) 2
   (d) 4

8. An mRNA has a codon with the base sequence GCA. What would the corresponding anticodon be?
   (a) ACG
   (b) AGC
   (c) GCA
   (d) CGU
   (e) CGT

9. If an error is made in the transcription of mRNA from DNA,
   (a) this constitutes a mutation that will be passed on to the next generation;
   (b) the tRNA will not know where to bind to the mRNA;
   (c) a protein will be formed which may have a different amino acid sequence from the normal gene product;
   (d) the mRNA will go to the wrong ribosome.

10. Key terms:

    semi-conservative replication   codon
    DNA polymerase   anticodon
    transcription   translation
    RNA polymerase   genetic code
    messenger RNA   triplets
    transfer RNA   hemoglobin
    ribosomal RNA   sickle-cell hemoglobin

SUGGESTED READINGS

Brown, D. D. The isolation of genes. Scientific American 229:21–29, 1973.

Darnell, J. E., W. R. Jelinek, and G. R. Molloy. Biogenesis of mRNA: genetic regulation in mammalian cells. Science 181:1215–1221, 1973.

DuPraw, E. J. DNA and Chromosomes. New York: Holt, Rinehart & Winston, Inc., 1970.

Miller, O. L., Jr. The visualization of genes in action. Scientific American 228:34–42, 1973.

Schekman, R., A. Weiner, and A. Kornberg. Multienzyme systems of DNA replication. Science 186:987–993, 1974.

Watson, J. D. Molecular Biology of the Gene. Ed. 3. Reading, Mass.: W. A. Benjamin Inc., 1976.

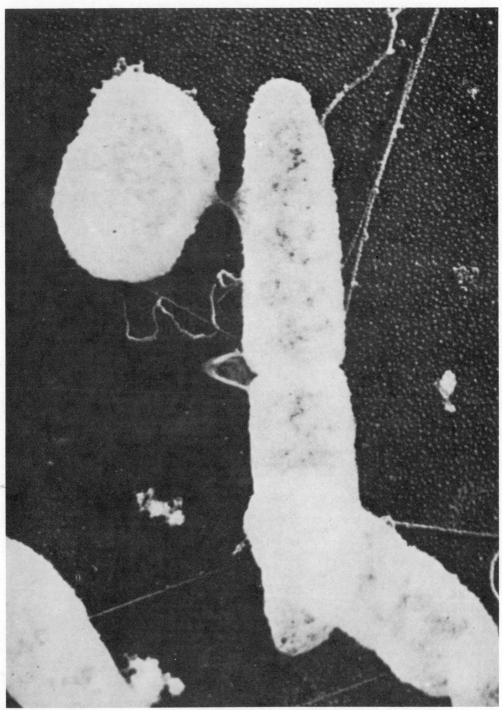

The transfer of genetic material from a donor bacterium *(right)* to a recipient cell across a short temporary bridge. This relatively rare phenomenon is regulated by a complex interaction of genetic factors. (Electron micrograph, shown at a magnification of 100,000×, by Thomas F. Anderson of the Institute for Cancer Research, Philadelphia.)

Chapter Nine

REGULATION

Most genes, like electrical appliances in the face of an energy shortage, are turned on only when there is a need for them. The rest of the time the genes are turned off, clearly an energy-saving device for the cell. There is indirect evidence that in mice only about three per cent of the ordinary genes in the liver transcribe RNA (are turned on), whereas about nine per cent of the genes in the brain transcribe RNA, that is, are activated for protein synthesis. Thus, we can say that the brain seems to be genetically busier than the liver.

The control of gene activity has now been widely studied for almost two decades. Much information about the process has emerged, but not enough to give us a detailed picture for humans and other complex organisms. Most of the studies and findings pertain to bacteria. Findings about the controls of gene activity in bacteria and other prokaryotes—the simple organisms which have no nuclear membrane and no chromosomal proteins—cannot be extended easily to higher plants and animals (the eukaryotes, which have cells with true nuclei). Nevertheless, we will look at the discoveries on genetic control in bacteria because they have served as a model for exploring control processes at work in higher organisms. The hope is eventually to understand how gene activities can be switched on and off at will, in order to correct inherited errors in gene activity that produce human defects.

REPRESSORS AND OPERONS IN BACTERIA

In bacteria, the work of F. Jacob, J. Monod, and their followers has given us the following description of genetic control. Structural genes, which code for ordinary proteins, are controlled by regulator genes which code for special proteins called repressors. For instance, E. coli cells have structural genes which determine specific enzymes for the utilization of lactose (the sugar in milk). One of these genes, the y gene, determines the enzyme that transports lactose molecules into the cell, and another, the z gene, breaks lactose down into two parts. (The products of yet other genes metabolize lactose further for use in cell nutrition.)

The z and y genes, and also the a gene (which we need not be concerned with) are grouped together on the bacterial chromosome (Fig. 9–1). This continguous group is called the lac operon. To one side of the operon is the i gene, which regulates z, y, and a. Between i and z on the chromosome lie two smaller sites, the promotor (p) and the operator (o). The promotor is the site where the protein called CAP can attach, and the operator is the site where repressor protein and also RNA polymerase can attach (but not at the same time).

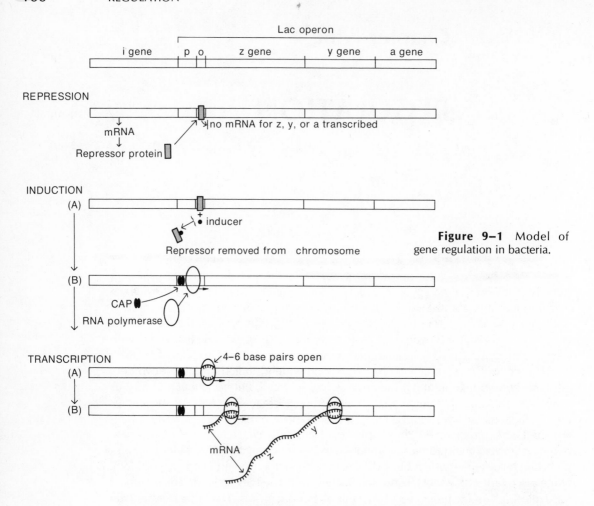

Figure 9–1 Model of gene regulation in bacteria.

When lactose is not present in the environment of a bacterium, the i gene product, the repressor, sits on the operator site (switch); this prevents the attachment of polymerase there. No mRNA is then transcribed from genes z, y, and a (Fig. 9–1). It is a clear advantage to the bacterium not to make digestive enzymes for nutrients which are not currently available.

When lactose is present in the environment, it (or a similar molecule) induces the synthesis of the lac enzymes. It combines with the repressor sitting on the operator, thereby moving the repressor for the "switch." This operator site can then be filled with RNA polymerase if the site has been prepared, in some as yet unknown way, by the previous attachment of the CAP protein to the promotor. The polymerase on the chromosome then glides along the operon and transcribes a z-y-a multigenic mRNA. Subsequently, at the ribosomes, three separate proteins are translated off of this multigenic mRNA.

How do these proteins bind to the promotor and operator switches of the lac operon? If one thinks of DNA base sequences as a bewildering succession of base pairs, as many geneticists did until recently, it is hard to visualize how particular proteins might recognize individual attachment sites on the DNA. Providentially, it is now known that switch sites have special properties in their DNA base sequence. Work in the laboratory of John Abelson and elsewhere indicates that both the CAP attachment site in the lac promotor and the repressor site in the operator have symmetrical properties (Fig. 9–2). The function of this twofold symmetry in these

The symmetrical DNA base sequence for CAP
attachment in the lac promoter reads:

The symmetrical sequence for repressor
attachment in the lac operator reads:

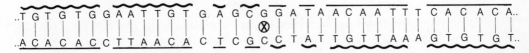

Ⓧindicates the pivot point in the sequence.

Note: looping in these symmetrical areas is
theoretically possible
but has not been
observed, at least in
the case of repressor
attaching to operator.

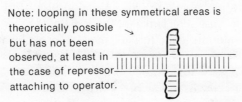

Figure 9–2   Symmetry in DNA.

sites is unknown. Although such areas could form complementary loops in the DNA (Fig. 9–2), no loops have been observed so far. Symmetrical regions in bacterial and viral DNA are also known, however, for other repressor-attachment sites, as well as for nuclease-attachment sites. (Nucleases are enzymes that break down nucleic acids.) All that can be said now about these switch sites is that the findings so far are puzzling.

In bacterial cells the level of enzymes coded by repressor-regulated genes varies, depending on whether an inducer is present or not. The level of enzymes in human cells also varies with time, differs in different organs, and responds to environmental inducers (such as hormones, as discussed subsequently). But whether or not enzyme levels in human cells are regulated by i-type genes, as in the lac operon, is disputed. Most workers in the field of gene regulation think the case for i-type genes in higher organisms has not been proved. Let us now consider the kinds of gene regulation that have been well documented for humans and other animals.

GENE ACTIVITY AND FROG DEVELOPMENT

Since frog eggs are larger and easier to manipulate than mammalian eggs, their nuclei can be removed by simple manipulations, thereby producing enucleated eggs. Briggs and King, Gurdon, and others have shown that frog nuclei can also be isolated under the microscope from tadpole intestinal cells or from cells of embryonic stages. When these nuclei are injected into enucleated frog eggs, the eggs develop into normal tadpoles (and adult frogs). The results are interpreted as meaning that, although many genes are "switched off" as cells mature and mass into specialized tissues, this gene inactivity is reversible, at least until fairly far along in frog development.

Precisely how these genes are repressed or derepressed in the frog is unknown. The studies do, however, serve as a comforting demonstration that genes are not simply lost or cast away as organs develop. Body cells retain their full potency in terms of genetic activity.

X-INACTIVATION

Another kind of evidence for differential activity of genes comes from the work of M. Lyon on mice and other mammals. In cats, for instance, tortoise shell coat color (rust and black) is virtually restricted to females. These coat color effects are due to X-linked alleles, tortoise shell females being heterozygous, $X^R X^{R'}$. The alleles R (rust) and R' (black) are expressed in separate patches on the cat's coat. XY males are either $X^R Y$ (rust) or $X^{R'} Y$ (black), but XXY males can be tortoise shell. (Tortoise shell males were avidly sought for studies of their chromosomes when it was first suspected they were probably XXY; the suspicion proved to be correct.) Now the question about the tortoise shell females (XX) or males (XXY) is: why are they patchy? Why aren't they uniformly dark rust all over? The explanation, called the Lyon hypothesis of X-chromosome inactivation, is that here in cats, and for a number of other X-linked cases in mammals, one of the two X chromosomes in every XX or XXY cell is inactivated (or repressed). A cell can thus express the X^R alone or the $X^{R'}$ alone, but not both.

X-chromosome inactivation occurs at random in cells in early embryonic life, when the embryo consists of as few as 60 cells. The descendants of an early XX cell with X^R-inactivated will form a patch of black on the skin later in development. Another cell, with $X^{R'}$-inactivated, will form a rust-colored patch.

For humans and other mammals, the cytological correlate of X-inactivation is the Barr body mentioned in Chapter Six. XX females have one Barr body, which is now known to be their inactive X chromosome: but different X's will be the Barr body in different cells. XXY males, of course, also have one Barr body and one inactive X. XXX females have two Barr bodies and two inactive X's.

One practical problem arising from X-inactivation involves the X-linked human gene called glucose-6-phosphate dehydrogenase (G6PD). Males with a deficiency of the G6PD enzyme suffer severe attacks of anemia when they consume certain foods (fava beans) or drugs (anti-malarial drugs). Females who are heterozygous for G6PD (deficiency/normal enzyme) are usually phenotypically normal; but a few heterozygous females have many red blood cells lacking the enzyme owing to inactivation of the X chromosome which bears the normal allele, and they approach the male G6PD deficiency phenotype. Heterozygosity for a harmful recessive allele does not always provide protection from genetic problems!

HORMONES

A well-documented example of hormonal effects on gene activity comes from the work of B. O'Malley and coworkers. Their studies involve the cells lining the

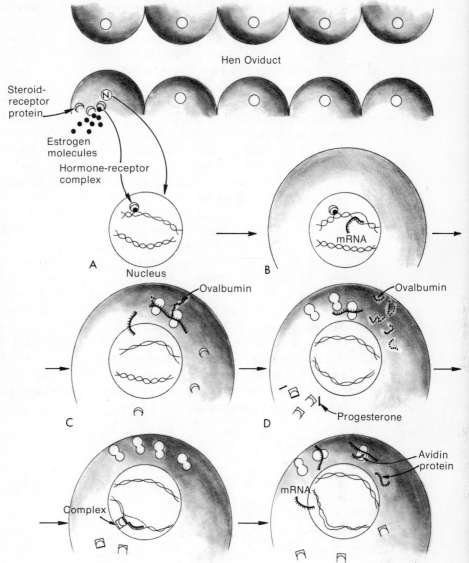

Figure 9–3 Hormone regulation of gene activity in hen oviduct. A: Hormone-receptor complex binds to chromosome. B: mRNA synthesis for ovalbumin is initiated. C: Ovalbumin is synthesized on ribosomes. D: Progesterone next invades cell and forms hormone-receptor complex. E: The complex enters the nucleus and binds to a chromosome. F: mRNA is synthesized and avidin is produced on ribosomes.

hen ovidict (through which the egg passes and where it picks up certain materials). One female hormone (estrogen) triggers these cells to synthesize the egg-white protein ovalbumin, and then another female hormone (progesterone) triggers them to synthesize a second egg-white protein, avidin. Both hormones activate the synthesis of the relevant mRNA's. Within two minutes after the administration of estrogen, mRNA synthesis is increased 40 per cent. These hormonal effects are quite specific in the sequence of activation of target genes in this tissue (Fig. 9–3). Although the hormones are small molecules which can diffuse into and out of all kinds of cells in the body, they affect only target cells which contain specific receptor proteins.

When the hormones enter the cell, they bind to the receptor proteins in the cytoplasm. The hormone-receptor complex diffuses into the nucleus and binds to chromosomal material (DNA). In some unknown way specific mRNA transcription is then initiated. Treatment with either hormone also enhances rRNA and tRNA synthesis and increases the population of ribosomes in the cytoplasm. In addition, estrogen may specifically activate the synthesis of the progesterone receptor.

Experiments in which isolated chick chromosomes were dissociated into DNA and protein fractions in vitro indicate that the so-called acidic proteins of chromosomes from oviduct cells, but not from cells of other tissues, facilitate binding of the hormone-receptor complexes to chromosomes. A number of questions remain unanswered about this process. For example, does the hormone-receptor complex bind to a promotor region on the chromosome? Does the complex activate the RNA polymerase system? Or does it work by some other means?

In humans, diffuse hormonal effects orchestrate the activity of a multitude of genes both simultaneously and sequentially. This is the situation, for instance, in the unfolding of the sexual phenotypes during development, in which many genes are involved. But untangling which hormones act on which human genes for "fine tuning" is still an elusive problem. For example, the menstrual cycle is dependent on an increase in estrogen, which triggers ovulation (the release of the egg from the ovary). Ovulation is followed by an increase in progesterone (Fig. 9–4). Here, as in the chicken studies, both estrogen-specific and progesterone-specific receptor proteins have been identified in the target tissue (uterus). But the genes in the uterine cells which are affected after the hormone-receptor complex enters the nucleus and binds to the DNA have not yet been identified. It is known, however, that there is only a limited number of binding sites on the DNA for the hormone-receptor complex. In human males, sex hormones also interact specifically with target cells (in the prostate gland, for instance) and become localized in cell nuclei, bound via a receptor protein to the DNA.

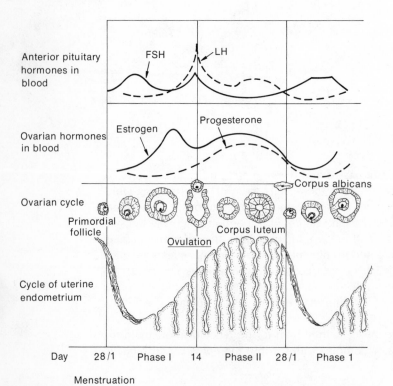

Figure 9–4 The time sequence of events in the menstrual cycle. Hormones in the hypothalamus area of the brain affect blood levels of pituitary hormones, which in turn act on the ovary. (From Clark, M.: Contemporary Biology. Philadelphia: W. B. Saunders Company, 1973.)

CHROMOSOMAL PROTEINS

As mentioned in the previous section, protein constituents of chicken chromosomes seem to enhance gene activity. Recent work has shown that of the two main kinds of chromosomal proteins in eukaryotes—histones and acidic proteins—the histones tend to depress gene activity generally and the acidic proteins tend to enhance it selectively. Since only about ten per cent of the genetic information in a cell is expressed at one time and since the histone:DNA ratio in eukaryotic chromosomes is about 1:1, histones clearly exist in the right amount and in the right place to fill the role of regulatory proteins for widespread gene repression.

Histones are proteins enriched in the basic amino acids lysine and arginine. The histone content of actively metabolizing cells and of inactive cells is about the same, a fact which does not, at first sight, conform to the idea of histones being major repressors. However, studies on cell-free biochemical systems have shown that adding histones to "naked" DNA blocks RNA transcription generally, and, conversely, that the partial removal of histones from chromosomal material (by treatment with dilute acid) enhances general RNA transcription.

Histones and DNA are synthesized simultaneously during the interphase period of the mitotic cycle (Chap. Four). When DNA synthesis is stopped experimentally by treatment with certain chemicals, histone synthesis (on cytoplasmic ribosomes) also stops. This finding suggests that as new DNA is synthesized histones quickly bind to it to repress unwanted RNA synthesis. Nevertheless, there is no indication that histones selectively repress the activity of particular genes. Evidence for *lack* of histone specificity includes the finding that histones are almost identical in amino acid composition in organisms as different as peas and cows.

The rest of the chromosomal proteins, other than histones, are called acidic or nonhistone proteins. Whereas there are only five main histones, there are hundreds of these acidic proteins. They differ greatly among themselves in size and in the length of time they stay associated with DNA in the nucleus before they are degraded. Acidic proteins isolated from chromosomal material contain a large number of enzymes, including the DNA and RNA polymerases. In contrast to the situation with histones, different species of plants and animals differ greatly in the kinds of nonhistones they possess. In addition, different tissues in the same organism differ in their nonhistone proteins, suggesting that these proteins are good candidates for specific gene regulator molecules. Evidence that the acidic proteins indeed *are* specific gene regulators includes the following:

(1) Acidic chromosomal proteins isolated from interphase cells (when RNA transcription is maximal) differ from those isolated from metaphase stages (when few genes are transcribing RNA).

(2) Estrogen treatment of cells has been observed to change the composition of the acidic proteins (but not the histones).

(3) Adding nonhistone proteins in vitro to a mixture of DNA and histones reverses the inhibition of RNA transcription caused by the histones.

(4) Fetal mouse liver synthesizes globin (the non-iron part of hemoglobin), but fetal mouse brain does not. Under experimental conditions, nonhistone proteins isolated from fetal liver, but not those from brain, lead to the production of globin-specific mRNA when they are mixed in vitro with DNA and histone from brain.

This evidence supports the hypothesis that nonhistone proteins dictate the transcription of particular genes. If this is true, how do they do it? A current theory by G. S. Stein and co-workers favors the idea that an individual nonhistone protein recognizes a certain site on the DNA, binds to it there, and then pulls off the generalized histone repressor from that site (Fig. 9–5). The histone removal may require the addition of phosphate groups to the nonhistone protein.

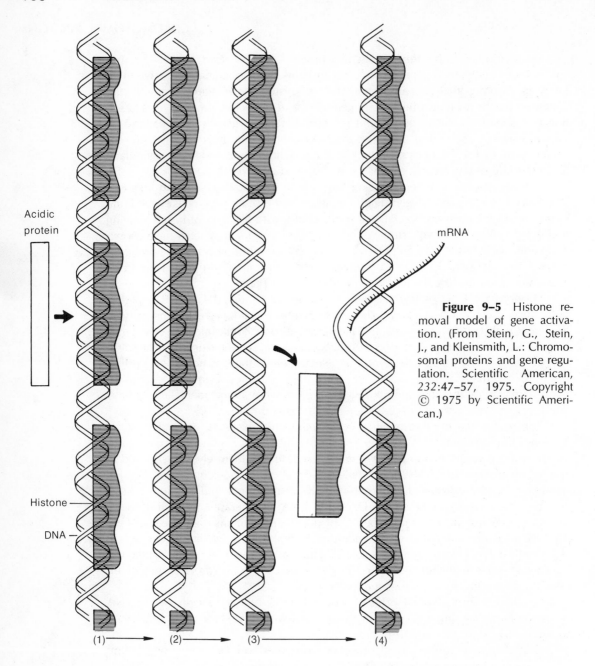

Acidic
protein

Histone

DNA

mRNA

Figure 9–5 Histone removal model of gene activation. (From Stein, G., Stein, J., and Kleinsmith, L.: Chromosomal proteins and gene regulation. Scientific American, 232:47–57, 1975. Copyright © 1975 by Scientific American.)

(1)⟶ (2)⟶ (3)⟶ (4)

SUMMARY

Although the broad outline of gene activity has been formulated, the regulation of this activity is not well understood. Various proteins seem to act as repressors or enhancers of DNA-directed RNA synthesis (transcription). Studies on simple organisms, the prokaryotes, have not given us a complete picture of the control of gene activity in complex organisms, the eukaryotes. Inactivation of the X chromosome, hormonal influences, and the disposition of chromosomal proteins on DNA all seem to be important to the regulation of eukaryotic genes in particular circumstances.

STUDY QUESTIONS

1. Suppose a mutation occurred in the i gene of bacteria such that it could no longer code for lac repressor. What effect would this have on the bacterial cell?

2. When nuclei isolated from cells of adult frogs are injected into enucleated frog eggs, the eggs do not develop beyond the embryonic stage. Comment.

3. A gene on chromosome No. 5 in the mouse is known to be the structural gene for the production of the enzyme glucoronidase in the kidneys. Genetic crosses among inbred lines of laboratory mice demonstrated that a second gene near the structural gene on the chromosome controls the induction of synthesis of glucoronidase by androgen (male hormone). How would you describe this second gene? (Based on the report of Swank and Bailey, 1973; see Suggested Readings for this chapter.)

4. Enzyme synthesis in the lac operon is controlled by:
   (a) DNA
   (b) histones
   (c) inducer, promotor
   (d) repressor, operator, inducer
   (e) promotor, operator, proteins, inducer
   (f) proteins

5. X-linked recessive alleles are expressed:
   (a) only in males
   (b) only in females
   (c) in both males and females
   (d) more frequently in males
   (e) more frequently in females
   (f) when on the X-inactivated homolog

6. The main components of eukaryotic chromosomes are:
   (a) DNA, RNA
   (b) DNA, uracil
   (c) DNA, polymerase
   (d) DNA, ribose, histone
   (e) DNA, RNA, nonhistone protein
   (f) DNA, RNA, protein

7. Key terms:

   prokaryotes   inducer
   eukaryotes   X-inactivation
   repressor   G6PD
   operon   hormone
   structural gene   hormone-receptor complex
   regulator gene   histones
   operator   acidic proteins
   promotor

8. Would you call the estrogen receptor a repressor molecule? Why?

SUGGESTED READINGS

Dickson, R. C., J. Abelson, W. M. Barnes, and W. S. Reznikoff. Genetic regulation: the lac control region. Science 187:27–35, 1975.

Gurdon, J. B. Transplanted nuclei and cell differentiation. Scientific American 219:24–35, 1968.

Hahn, W. E., and C. D. Laird. Transcription of nonrepeated DNA in mouse brain. Science 173:158–161, 1971.

Lewin, B. Number of cells at the time of X-activation. Nature 249:9–11, 1974.

Lyon, M. F. Chromosomal and subchromosomal inactivation. Annual Review of Genetics 2:31–52, 1968.

Markert, C. L., and H. Ursprung. Developmental Genetics. Englewood Cliffs, N.J.: Prentice-Hall, Inc., 1971.

O'Malley, B. W., and A. R. Means. Female steroid hormones and target cell nuclei. Science 183:610–620, 1974.

O'Malley, B. W., and W. T. Schrader. The receptors of steroid hormones. Scientific American 234:32–43, 1976.

Ptashne, M., and W. Gilbert. Genetic repressors. Scientific American 222:36–44, 1970.

Stein, G. S., J. S. Stein, and L. J. Kleinsmith. Chromosomal proteins and gene regulation. Scientific American 232:47–57, 1975.

Swank, R. T., and D. W. Bailey. Recombinant inbred lines: value in the genetic analysis of biochemical variants. Science 181:1249–1251, 1973.

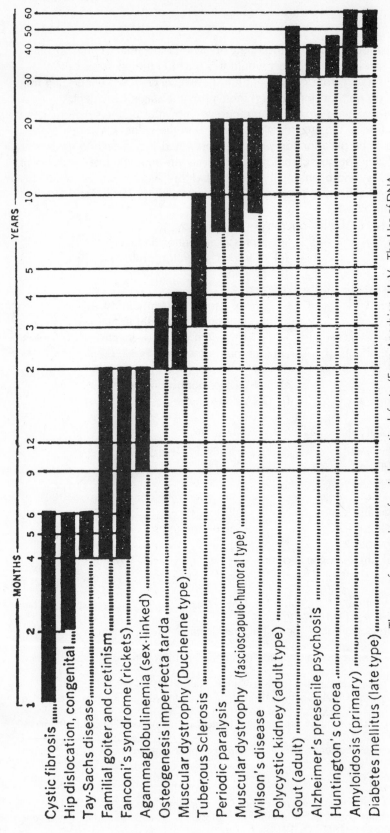

The age of expression of varied genetic defects. (From Aposhian, H. V., The Use of DNA for Gene Therapy – The Need, the Experimental Approach and Implications. Perspectives in Biology and Medicine, 14:98–108, 1970.)

Chapter Ten

GENES, METABOLISM, AND DEVELOPMENT

We have discussed DNA, RNA, and proteins because of the intrinsic elegance of this genetic triad. It shows the simpleness of nature's mechanisms and provides the biochemical unity of the living world. Knowledge of this triad provides a foundation for mounting experimental probes into many aspects of gene effects, including human disease.

INBORN ERRORS OF ENZYME FUNCTION

The first notice of a relationship between disease and genetics occurred shortly after the rediscovery of Mendel's work in 1900. A. E. Garrod, a British physician, studied several rare, human metabolic defects, including alkaptonuria and cystinuria. He pointed out in 1902 that such defects follow recessive inheritance patterns of Mendelian segregation, with the majority of cases showing up in the offspring from marriages between cousins. Some of Garrod's "inborn errors of metabolism," as he called them, fit the category of a disease, in the ordinary sense, and some do not. Cystinuria, for instance, is a harmful condition: the excessive excretion of cystine and several other amino acids in the urine eventually builds up painful deposits of materials in the urinary tissues. Alkaptonuria, on the other hand, is not harmful. Its most noticeable symptom is a darkening of the urine upon exposure to air. It is found in one of every 250,000 persons throughout the world. Garrod realized that many inherited defects in metabolism, in addition to alkaptonuria, cystinuria, and albinism (which he also studied), may exist. He wrote:

> May it not well be that there are other such chemical abnormalities which are attended by no obvious pecularities and which could only be revealed by chemical analysis? If such exist and are equally rare . . . they may well have wholly eluded notice up till now. . . . The thought naturally presents itself that these inborn traits are merely extreme examples of variations in chemical behavior which are probably everywhere present in minor degrees and that just as no two individuals of a species are absolutely identical in bodily structure neither are their chemical processes carried out on exactly the same lines.

We now know that Garrod's inborn errors are the results of enzyme defects produced by structural, or perhaps regulatory, genes. Albinism, as mentioned in Chapter Three, is the result of an enzymatic inability to perform one of the chemical steps necessary to convert the amino acid tyrosine into the colored pigment molecule melanin. This enzymatic defect may occur at any of several steps in the biochemical pathway between tyrosine and melanin (Fig. 10–1). That is, albinism is genetically heterogeneous. Several other inborn errors associated with enzymatic

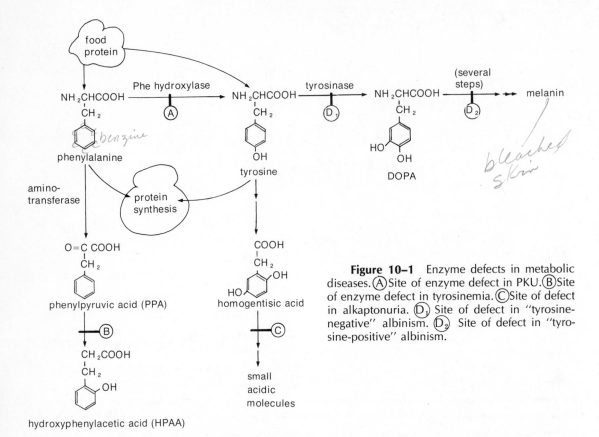

Figure 10–1   Enzyme defects in metabolic diseases. (A) Site of enzyme defect in PKU. (B) Site of enzyme defect in tyrosinemia. (C) Site of defect in alkaptonuria. (D₁) Site of defect in "tyrosine-negative" albinism. (D₂) Site of defect in "tyrosine-positive" albinism.

defects in the metabolism of tyrosine and the related amino acid, phenylalanine, are known and are also diagrammed in Fig. 10–1. These include alkaptonuria and phenylketonuria (PKU). The latter is another autosomal recessive trait. (See Box on Phenylketonuria, p. 111.) It can be described as a full-blown chronic disease because it causes serious mental retardation—although its effects can be subdued if treatment is started soon after birth. Hundreds of inborn errors, or metabolic diseases, are now known in the human population. Similar metabolic diseases have been discovered in mice, dogs, and other animals studied in the laboratory.

PHENYLKETONURIA (PKU)

Phenylketonuria is a disease more common to populations of northern European ancestry than any others. Yet even in Ireland, the country with the highest known frequency, it occurs only once in every 4500 births. It is an autosomal recessive, congenital disease. Among U.S. whites, the frequency of heterozygotes—PKU infants have parents who are necessarily heterozygotes—is about one in 50, and the incidence of PKU is about one in 14,000 births. The only genetic disease among U.S. whites with a higher incidence than PKU is cystic fibrosis; it occurs once in every 2,500 births.

Because PKU individuals lack a functional enzyme (phenylalanine hydrolase) for converting phenylalanine from food proteins or other metabolic sources into tyrosine, they have a high level of phenylalanine in the blood and a high level of modified compounds—phenylpyruvic acid (PPA) and hydroxyphenylacetic acid (HPAA) — in the urine. PKU is thus a chronic disease in which normal chemicals are present in excess amounts. The PKU phenotype is defined as more than 20 milligrams (mg) of phenylalanine per 100 milliliters (ml) of blood serum in conjunction with the presence of PPA and HPAA in the urine under normal dietary conditions. This definition precludes misclassifying those persons with transient high levels of phenylalanine in the blood (from other causes) as having PKU. If untreated, PKU patients have lasting physical and mental abnormalities (retardation) and a shortened life span; they are bleached-looking in their skin, hair, and eye color owing to the impairment of melanin production (Fig. 10–1); and they often have convulsions and abnormal electro-encephalogram (EEG) patterns. The cause of the brain damage in PKU children is still unknown. Intermediate levels of phenylalanine in the blood serum (2 to 12 mg per 100 ml), caused by other factors, do not produce retardation and related problems.

Phenylketonuria was identified in 1934. By 1953 it was shown that a low-phenylalanine diet alleviates the mental and physical effects of the PKU phenotype. (Some phenylalanine is needed in the diet by everyone, of course, to synthesize proteins; but over half the amount of phenylalanine taken into the body from food is normally converted into tyrosine by the enzymatic route shown in Fig. 10–1.)

There is now a cheap, easy test (the "diaper test") for detecting PKU in newborns. An iron chloride compound ($FeCl_3$) reacts with phenylpyruvic acid in PKU urine to produce a green color. This or another screening test is now required by law for all births in most states in the U.S. Once PKU is detected, treatment consisting of a low-phenylalanine diet (partly in the form of a synthetic food formula) is immediately started, preferably before the infant is one month old. This treatment partially, but not fully, counteracts the harm of the enzyme deficiency. L. F. Saugstad has reported that mothers of PKU infants have a high rate of "obstetric complications;" evidently prenatal damage to these PKU infants often results in neonatal jaundice, epileptic seizures, and retarded growth. How long the low-phenylalanine diet should be continued for PKU children is not yet established, but the current guess is at least until 6 years of age (at which time the brain is about 90% of adult size).

Ferric Chloride

The success of this dietary treatment depends greatly on the understanding and cooperation of the PKU child's family. Genetic counseling clinics (Chap. Fifteen) offer dietetic, genetic, and sociological counseling for PKU families. PKU heterozygotes cannot yet be reliably identified by a simple chemical test. Most of them are identified only when they become the parents of a PKU child. It is hoped that research now in progress will lead to a test for easily identifying PKU heterozygotes in the near future.

BIOCHEMICAL PATHWAYS

The effect of an enzyme deficiency is to cause a block somewhere in the steps for the conversion of food molecules into useful materials for the cell or in the steps for their later breakdown into waste products. As is true for PKU, any block in just one biochemical pathway will have several adverse chemical effects. For instance, one effect of PKU is a reduction of the amount of melanin produced by the body, since melanin results from conversions occurring after the block in the phenylalanine pathway. This situation is known as an end-product deficit effect. A second effect is the accumulation of substances in the path before the block: for PKU it is phenylalanine that accumulates. A third possible effect is an "overflow," or shunt, of the accumulated substance into a side pathway: for PKU this produces an excess of secondary products in the blood, urine, and brain (where it causes retardation).

Although PKU can be treated by a low-phenylalanine diet, another recessive metabolic disease, Tay-Sachs disease (TSD), cannot. This disorder also results from an accumulation of a substance before an enzyme block, but dietary restriction for TSD individuals will not change the fatal course of the disease: the offending material (lipid) in the tissue does not arrive preformed in the diet, but is synthesized by body cells themselves from smaller molecules. (See Box on Tay-Sachs disease, below.

Enzymatic pathways in metabolism have been fairly thoroughly worked out, in large part by studies which have experimentally induced enzymatic defects in strains of microorganisms. The naturally-occurring enzyme defects in humans have also helped unravel the pathways and byways of metabolism. All indications are that cells of all organisms have the same basic enzymatic pathways for breaking down and synthesizing biological molecules. But since most enzyme blocks are lethal at an early stage of development, they are found only rarely in human populations or among plants and animals in the wild. However heritable enzyme defects induced in laboratory animals can be studied by researchers on a long-term basis by the clever manipulation of diet and of other conditions of growth.

TAY-SACHS DISEASE (TSD)

The brain contains many membranes in and around its cells. These membranes contain the fatty substances called lipids. When lipids are not broken down properly during cell metabolism, they build up in cells. The build-up results from a deficiency of one or another of the enzymes involved in lipid metabolism. Excess lipid conditions are called lipid storage diseases. Ten such storage diseases have been identified, each due to a characteristic enzyme deficiency. Many of these diseases involve mental deterioration because of the excess lipid in brain tissues. One such disease, Tay-Sachs disease, is an autosomal recessive condition. It occurs in one out of 6000 births in marriages between Jews from northern Europe, particularly Poland, Lithuania, and Russia, but in only one out of 600,000 births in marriages between others of European descent. About one in 40 persons of European Jewish ancestry is a carrier for the disease.

TSD infants are normal at birth but become progessively retarded in muscle coordination, progressively blind and listless, and prone to convulsions. They die by the age of four years. The brain at that time is composed of cells engorged with excess lipid bodies (Fig. 10–2).

As a result of recent research, the TSD condition can be diagnosed by a deficiency of the enzyme hexosaminidase A. The hex A enzyme test can even be done in utero, in time to allow for therapeutic abortion if the parents desire it (Chap. Six). Furthermore, the enzyme test can detect

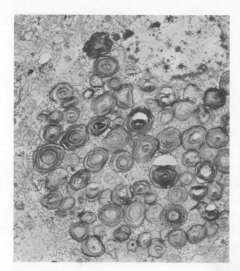
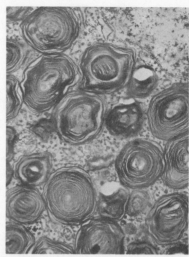
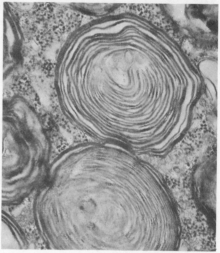

Figure 10–2 Three levels of magnification of brain cells with excess lipid in the TSD condition. (Courtesy of John S. O'Brien, University of California, San Diego. Reprinted from Advances in Human Genetics, Vol. 3, 1972, p. 45. Courtesy of Plenum Publishing Corporation.)

TSD carriers; they have an enzyme level intermediate between those of TSD homozygotes and normal individuals. The hex A test is the basis for the current genetic screening programs for TSD carriers in many Jewish communities in the U.S.

Treatment for TSD is nonexistent. One possibility involves administration of the missing hex A enzyme by injection into the bloodstream. This requires a ready supply of the enzyme and the ability of the body to transport the injected enzyme to the target tissue, which for TSD is the brain. Hex A enzyme can be isolated from normal placenta and urine; but when it is injected, very little manages to enter the brain. The obstacle here is the "blood-brain barrier," a poorly understood but vital device which protects the brain from many toxic and foreign substances (although, of course, not from hallucinogens). An early breakthrough in TSD treatment currently seems unlikely.

A recent report that TSD heterozygotes may be genetically resistant to tuberculosis might help explain the maintenance of the Tay-Sachs allele in the historically European branch of the Jewish population. It suggests that the lethal effect of TSD homozygosity may have been offset by the advantage of the TSD heterozygote in fending off tuberculosis. Tuberculosis has been in Europe for a long time, although of course not confined to Jewish populations there.

TIMING AND LOCATION OF GENE ACTION

Unlike the gene involved in TSD (which can be detected by amniocentesis in utero), some genes are not active all the time, but only at certain stages of development. There are many known examples of genes which act at specific times in humans and in other species. We have already seen how sex hormones act on the genes of target tissues at assigned times (Chap. Nine). And the liver enzyme (phenylalanine hydrolase) missing in PKU individuals is also a "timed" gene product. It is first detectable at 11 to 12 weeks after conception and reaches its maximal (adult) level by birth. Other enzymes are known which do not reach their maximal level until several weeks or months after birth—for example, the aminotransferase enzyme that converts phenylalanine to PPA (Fig. 10–1).

Hemoglobins: Synthesis and Development

The protein hemoglobin has a particularly interesting pattern of developmental changes. These changes are intimately related to an understanding of several, but not all, hemoglobin-defective human diseases. Since hemoglobin is not an enzyme, it is not involved in a metabolic pathway; thus inherited hemoglobin defects have no symptoms that result from accumulated chemical side products. The situation is more straightforward: a deficit of hemoglobin means less oxygen going to all the tissues and, consequently, less efficiency in metabolic and physiological processes generally.

Hemoglobin binds oxygen when oxygen is in high concentration and detaches it when the environment has little of it. This property allows hemoglobin to take up oxygen in the lungs and release it to the body tissues, where it is used in metabolism. During pregnancy, however, the oxygen that arrives at the placental barrier between the mother's and fetus's circulation is already at a lower concentration than when it was in the mother's lungs. It is not surprising then, that a different kind of hemoglobin functions in the fetus, one that can pick up oxygen even when it is at low concentration. This hemoglobin is called fetal hemoglobin (Hb F), and is composed of two alpha (α) and two gamma (γ) protein chains (denoted $\alpha_2\gamma_2$). In contrast, adult hemoglobin (Hb A) is composed of two α and two beta (β) chains ($\alpha_2\beta_2$). The switch from the synthesis of γ chains to β chains—another example of timed gene activity—occurs just before birth (Fig. 10–3). This timed switch does not occur, however, in those individuals, known as β-thalassemics, who are genetically unable to synthesize β chains. They therefore lack normal hemoglobin A and are chronically anemic.

Beta-thalassemia is found from the Mediterranean to India and Southeast Asia. In β-thalassemic individuals fetal hemoglobin production continues until the time of death, which is usually before puberty. Beta-thalassemia is an autosomal recessive disease. The parents of a β-thalassemic child are necessarily carriers of the inherited anemia. They have only one functional β-chain allele, the other (β^{thal}) being inactive. They sometimes have a modicum of Hb F in their blood as a result

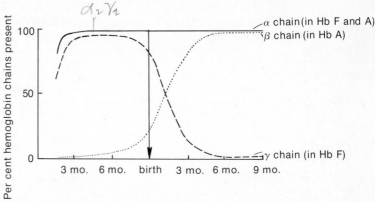

Figure 10–3 Developmental changes in human hemoglobin.

of incomplete switching of hemoglobin genes at birth. Clearly in the case of β-thalassemia, the γ-chain genes are permanently active (derepressed) and the β-chain genes permanently repressed or deleted. So in this disease an error in regulation of gene activity rather than an amino acid change in a gene product is at least partially at fault. The other kind of hemoglobin error we have discussed, sickle-cell anemia (Chap. Eight), does however result from an amino acid change. (See Table 10–1.)

Isozymes and Development

We have seen in the case of hemoglobin how gene activity can differ with time of development. In addition, gene activity can differ by location in multicellular organisms, some tissues exhibiting one kind of activity and other tissues another. For example, hemoglobin is synthesized only in red blood cells, and phenylalanine hydrolase (the PKU enzyme) is found in the liver and to a lesser extent in the kidney, but not in muscle, lung, or brain.

The classic example of differential gene-enzyme activity in different tissues involves the enzyme lactate dehydrogenase (LDH). It plays a key role in the breakdown of sugars (carbohydrate metabolism). C. Markert and his co-workers have shown that LDH is present in virtually all tissues in the body; but when tissue extracts are analyzed by the method of electrophoresis (Fig. 10–4), different tissues are shown to have different kinds of LDH (that is, different LDH isozymes).

The findings suggest that there are two separate genes involved in LDH formation, just as there are two genes which determine the protein subunits of hemoglobin A. The two LDH genes are called A and B. A and B here refer to separate DNA sequences, not to alternative alleles of one sequence. Genes A and B code for LDH protein chains (subunits), which are also called A and B, respectively. These subunits can combine at random in groups of four in the cytoplasm to make the finished protein (the tetrameric or four-part LDH molecule). The five possible combinations for tetramer formation are: AAAA, AAAB, AABB, ABBB, and BBBB. All five have been isolated from various human tissues, but different combinations predominante in certain locations (Fig. 10–4). The BBBB form predominates in heart and kidney, while the AAAA form predominates in liver and skeletal muscle. For heart tissue the ratio of A to B subunits is about 1:20, for kidney 1:10, and for muscle 10:1. The A and B genes clearly have independent levels of activity in different tissues.

A few individuals have been found who are heterozygous for mutant alleles at either the A or B gene for LDH. These individuals, and the even rarer mutant homozygotes, have no known medical abnormalities. Presumably these mutations cause innocuous amino acid substitutions in the LDH protein subunits.

TABLE 10–1   Review of Some Inherited Anemias*

Genotype	Phenotype	Location
HbS/HbS	Sickle-cell anemia (red cells deformed into sickle shape; hemoglobin all of type $\alpha_2^A\beta_2^S$ due to specific amino acid change in beta chain)	Parts of Africa and Middle East; U.S. and other areas with persons of African, Greek, and other ancestry
HbA/HbS	Sickle-cell carrier (red cells normal in shape; hemoglobin of two types: $\alpha_2^A\beta_2^A$ and $\alpha_2^A\beta_2^S$; no anemia; resistant to malaria; see Chap. 17)	As above
HbC/HbC	Hemoglobin C anemia (less severe than sickle cell; hemoglobin all $\alpha_2^A\beta_2^C$ due to a specific amino acid change in beta chain)	Primarily West Africa
HbA/HbC	Hemoglobin C carrier (no anemia)	As above
Hbβthal/Hbβthal	Beta-thalassemia (severe anemia; no $\alpha_2^A\beta_2^A$ but $\alpha_2^A\gamma_2^A$ and other kinds of hemoglobin present)	Mediterranean area primarily, but also through Asia to Southeast region
HbA/Hbβthal	Thalassemia carrier (probably resistant to malaria)	As above
HbA/HbA	Normal ($\alpha_2^A\beta_2^A$)	Worldwide
G6PD⁻/Y or G6PD⁻/G6PD⁻	X-linked anemia due to enzyme deficiency (Chap. 9); may be resistant to malaria; see Chap. 19)	Primarily Old World tropics
G6PD⁻/G6PD⁺	Female carriers of anemia	As above
G6PD⁺/Y or ⁺/⁺	Normal enzyme levels	Worldwide

*Some anemias are due to a deficiency of iron or other environmental factors; others are due to genetic factors.

Abbreviations: Hb = hemoglobin; S = sickle-cell hemoglobin allele; A = normal hemoglobin allele; C = another hemoglobin allele; thal = thalassemia; G6PD⁻ = glucose-6-phosphate dehydrogenase deficiency; Y = Y chromosome.

The situation for certain isozyme variants in muscular dystrophy may not be innocuous, however. Muscular dystrophy (Duchenne type) is an X-linked recessive disease. Affected males have muscle degeneration which makes them unable to walk by the age of 10 years and usually proves fatal by the age of 20. Heterozygous females are not affected. The disease has variously been ascribed to a defect in the nerves attached to skeletal muscles, a defect in cellular membranes of muscle cells, and a defect in glucose metabolism in muscles. Among other findings, it is clear that muscle tissue from muscular dystrophy patients has a reduced ability to metabolize glucose normally; instead, their muscle tissue shunts glucose to a side pathway (recall the shunt problems in PKU, discussed earlier). Since normal glucose metabolism is needed to produce adequate amounts of ATP for muscle cell maintenance and movement, a defect in an enzyme in the glucose pathway is understandably injurious. A recent report indicates that the enzyme hexokinase

Figure 10–4 Lactic dehydrogenase forms in human tissues. Tissue extracts are applied to a jelled medium and subjected to an electric current (marked by arrow). Proteins which differ in net electrical charge are thus separated. The positions of the proteins in the gel at the end of the experiment are detected with a specific dye. This process is called gel electrophoresis. (After Harris, H.: The Principles of Human Biochemical Genetics. London: North Holland Publishing Company, 1970, p. 40.)

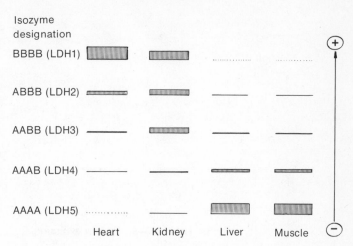

(the first enzyme in the pathway of glucose breakdown) occurs in various isozyme forms in different human tissues, and that one of the hexokinase isozymes in normal muscle is scarcer than that in muscular dystrophy patients. The normal isozyme also differs in electrophoretic mobility. Other causes of muscle degeneration, such as alcoholism and virus infections, do not cause an alteration in hexokinase.

What then can be said about the significance of isozymes? Many enzymes are now known to exhibit different isozyme forms in many organisms. How isozymes facilitate the specialization of tissues during development, and, moreover, how the relevant genes are switched on or off in different tissues, are questions still under investigation. Perhaps the subject of isozymes is as good a reason as any to mention that science seems to be an endless puzzle.

EVOLUTION OF PROTEINS

The β chain and the γ chain of human hemoglobin both have 146 amino acids. They are identical in 107 of these 146 amino acids. The α chain of human hemoglobin has 141 amino acids, 61 of which are identical with those in the β chain. Let us consider this group of facts in conjunction with the known multiplicity of LDH genes (regarding which, although we do not know the amino acid sequences involved, we can infer a resemblance because they fulfill the same enzymatic function). A question that emerges is: why does all this similarity and multiplicity exist? Biologists are prone to think about this question in terms of ancestral genes that, in the course of time, have by chance multiplied (producing duplicate genes) and then selectively diverged by subsequent differential mutations which permit specialized adaptations.

Since more is known about the hemoglobin molecule than any other protein, it will serve well as an illustration of protein changes through eons of time.

Hemoglobin A is identical in humans and chimpanzees. In fact, the average human protein is over 99 per cent identical with its chimp counterpart (Chap. Nineteen). But species more distant in evolutionary relationship have proteins correspondingly different from those of humans (Fig. 10–5). Whereas the α chain of hemoglobin, for example, differs by zero amino acids between humans and chimp, it differs by two amino acids between humans and gorilla, by 18 amino acids between humans and pig, and by 27 amino acids between humans and rabbit. These findings may help to explain why in the biological realm it is difficult to think about evolution without also thinking about its DNA-controlled genetic basis, just as it

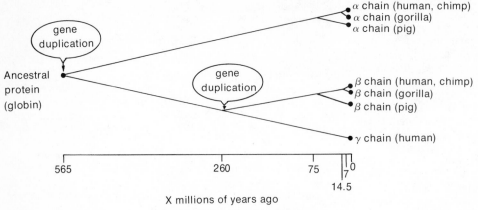

Figure 10–5 Protein evolution.

is difficult to think about genetic products without thinking about their functions in evolutionary time. The adaptiveness of genetic biochemistry enfolds us all.

SUMMARY

By looking at errors in gene activity we have seen how genes can cause inherited diseases, and how they affect the normal course of chemical conversions at the cellular and whole-body levels. For an inherited recessive disease, knowing the genetic lesion involved and its effects is invaluable for the design of medical treatment and for identifying carriers. Medical treatment of metabolic disease, however, rarely prevents the development of the abnormal phenotype and may be required throughout the lifetime of the affected individual. By the study of a large number of inherited chemical defects it is now becoming possible to approach the question of adaptiveness and maladaptiveness in our inherited chemical makeup.

STUDY QUESTIONS

1. The homozygous condition known as alpha-thalassemia (absence of the alpha chain of hemoglobin) causes spontaneous abortion of the fetus. Discuss possible reasons for this.
2. If two isozyme subunits differ in a certain number (X) of amino acid residues, does this mean their DNA sequences also differ in X number of base pairs?
3. Comment on the following: "Proteins are living fossils of evolutionary history."
4. Myoglobin is the oxygen-binding protein found in muscle. It is a single-chain molecule 153 amino acids long. Human myoglobin is more like whale myoglobin in amino acid sequence than it is like either the α or β chains of human hemoglobin, although it has similarities to the latter two molecules. Where would you put the ancestral myoglobin in relation to the globin evolutionary sequence shown in Figure 10–4?
5. Discuss possible evolutionary advantages of duplicate genes, in general, for an organism.
6. Albinism actually comes in two forms (contrary to the discussion of a single cause given in Chap. Three). The two forms are called tyrosinase-negative and tyrosinase-positive. Tyrosinase is an early enzyme in the pathway which converts the amino acid tyrosine into the pigment molecule melanin in nonalbinos (Fig. 10–1). Tyrosinase-positive albinos are usually less white than the others, having a yellowish tinge to the hair, skin, and iris of the eye. The presence of tyrosinase

can be demonstrated in the yellowish albinos and in nonalbinos by a simple "hair test." A hair with its attached root is pulled from the head. The root contains enzymes; if the root contains tyrosinase, it will turn dark (i.e., form melanin) when incubated in a concentrated solution of tyrosine for many hours in a test tube. By this test the yellowish albinos have been shown to be tyrosinase-positive. This finding may eventually lead to treatment for this form of albinism; however, feeding solutions of tyrosine to yellowish albinos has been tried but has not proved therapeutic.

The two forms of albinism are known to be nonallelic; several normally pigmented children of parents who are "positive" and "negative" for tyrosinase have been recorded. If a is the allele for albinism due to lack of tyrosinase and b is the allele for albinism which is tyrosinase-positive, the parents of such normally pigmented children could be designated by which of the following genotypes?

(1) aa bb × AA bb   (3) Aa Bb × Aa Bb   (5) aa bb × aa bb
(2) aa BB × AA bb   (4) Aa bb × Aa Bb

7. Key terms:

inborn errors   beta-thalassemia
biochemical pathway   isozyme
genetic heterogeneity   LDH
phenylketonuria   electrophoresis
Tay-Sachs disease   muscular dystrophy
fetal hemoglobin   protein evolution

SUGGESTED READINGS

Brady, R. O. Hereditary fat-metabolism diseases. Scientific American 229:88–97, 1973. Discusses Tay-Sachs disease.

Culliton, B. J. Cooley's anemia: special treatment for another ethnic disease. Science 178: 590–593, 1972. Discusses the state of research on beta-thalassemia.

Garrod, A. E. The incidence of alkaptonuria: a study in chemical individuality. Lancet 2: 1616–1620, 1902.

Glass, R. D., and D. Doyle. Genetic control of lactate dehydrogenase expression in mammalian tissues. Science 176:180–181, 1972.

Goodman, M., G. W. Moore, and G. Matsuda. Darwinian evolution in the genealogy of hemoglobin. Nature 253:603–608, 1975.

Markert, C. L., J. B. Shaklee, and G. S. Whitt. Evolution of a gene. Science 189:102–114, 1975.

Myrianthopoulos, N. C., and S. M. Aronson. Population dynamics of Tay-Sachs disease. II. What confers the selective advantage upon the Jewish heterozygote? In Proceedings of the Fourth International Symposium on Sphingolipodoses. New York: Plenum Publishing Corp., 1973.

Nitowsky, H. M. The significance of screening for inborn errors of metabolism. In Heredity and Society. I. H. Porter and R. G. Shalko (eds.). New York: Academic Press, Inc., 1973, pp. 225–260.

Rosenberg, L. E., and C. R. Scriver. Disorders of amino acid metabolism. In Duncan's Diseases of Metabolism. Ed. 7. P. K. Bondy and L. E. Rosenberg (eds.). Philadelphia: W. B. Saunders Co., 1974, pp. 465–654.

Saugstad, L. F., The influence of obstetric complications on the chemical picture of classical phenylketonuria. Clinical Genetics 4:115–124, 1973.

Smith, D. W. Reticulocyte transfer RNA and hemoglobin synthesis. Science 190:529–534, 1975.

Strickland, J. M., and D. A. Ellis. Isozymes of hexokinase in human muscular dystrophy. Nature 253:464–466, 1975.

Zuckerkandl, E. The evolution of hemoglobin. Scientific American 212:110–118, 1965.

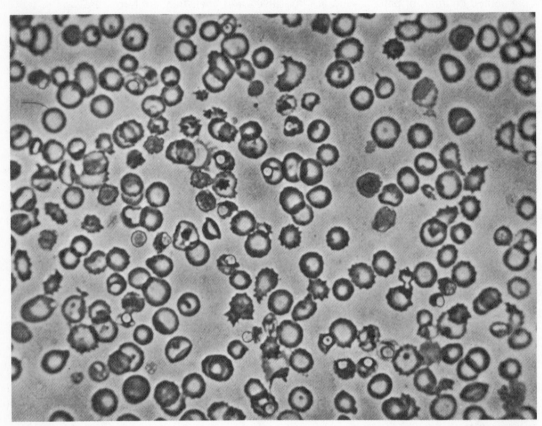

Blood from a baby born with erythroblastosis. The baby was Rh-positive and developed in an Rh-negative mother who had been sensitized by previously bearing an Rh-positive baby. Many of the red blood cells are deformed and do not carry oxygen well. A. M. Winchester: Heredity, Evolution, and Humankind. West Publ. Co., St. Paul. 1976.

Chapter Eleven

IMMUNOGENETICS

One of the frontiersmen of modern evolutionary theory, J. B. S. Haldane, pointed out that infectious disease probably was the major force of the past threatening the survival of humanity and selecting against the continuation of certain genotypes. Infectious diseases in the past have been kept under control somewhat by traditional remedies, exercise, prudent nutrition, sanitary practices, or a sparse population. Today we know that, in addition to such external defenses against disease, internal defenses are provided by the body, chiefly through the immune system.

Immunogenetics is a new field of research which studies the relationships between the genetic, chemical, and disease-preventing mechanisms of the immune system. The immune system is a topic with ramifications into problems of heart transplants, incompatibility between individuals in blood transfusions, the mechanisms of vaccination against smallpox, and the prevalence of allergies. The immune system defends the body against foreign invasions by toxic substances (insect bites), disease-causing bacteria, viruses, and other organisms. But the system may also be involved in protection against internal "invasions" by cancer cells. Because of such practical problems, the study of immunity at the molecular, cellular, and even population level is vigorously pursued in many laboratories today.

ANTIBODIES AND ANTIGENS

Human beings, along with their vertebrate relatives, have the ability to generate certain kinds of cells, lymphocytes, which produce and secrete certain protective proteins, antibodies. Antibodies circulate within the bloodstream and other body fluids and defend the body from disease-causing organisms. Lymphocytes and antibodies are the major components of the immune system.

Antibodies are large proteins. Many have the shape of a two-headed pair of pliers (Fig. 11–1) and are constructed of two identical heavy (long) and two identical light (short) amino acid chains. The four chains are held together by covalent bonds involving sulfur atoms (disulphide bonds). The central area of the antibody molecule is relatively constant in amino acid sequence. The tips of the two heads are, however, variable in sequence in both the heavy and light chains, thus forming differently structured pincer areas on the various kinds of antibodies within each individual. The different pincer areas give antibodies their specificity for reacting with particular foreign substances (antigens).

An individual has a multitude of types of antibodies. They are the products of

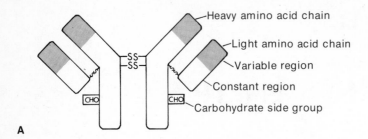

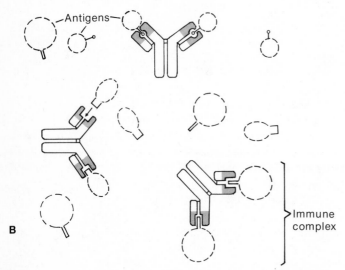

Figure 11–1   A: General shape of antibody molecule. B: Specific antigen–antibody reaction: "lock and key."

many different genes (which are thought to have arisen by repeated duplication of one original gene). A gene which specifies one kind of heavy chain has been identified; called Gm, it is known to have a large number of alleles. The gene called Inv specifies the light chain; it also has many alleles. The total number of genes specifying heavy and light chains is not known. In any event, once heavy and light chains are produced on ribosomes, they combine into functional antibodies. A single cell apparently produces only one kind of antibody molecule. Nevertheless, the body is able to protect itself against many kinds of foreign molecules because different genes coding for antibody chains are active and expressed in different cell lines, or clones —according to the popular "clonal selection theory" of acquired immunity (Fig. 11–2). These different cell lines may arise early in development as a result of somatic mutation in body cells, that is, owing to rare changes in the somatic cell's DNA. (See Chap. Thirteen for details of the mutational process.)

Antigens are proteins, carbohydrates, or other kinds of large molecules which are foreign to the body, that is, different from the body's own molecules. Some antigens are parts of the surface of infectious bacteria or other foreign cells that get into the body, whereas others are "free" antigens. Only a small patch on the antigenic molecular surface reacts with an antibody. If an individual has antibodies specific for an invading antigen, they will form a complex of one antibody molecule locked onto two antigen molecules. This lock-and-key concept for specific antibody-antigen interaction is diagrammed in Figure 11–1. The formation of this immune complex usually renders the antigen harmless to the body. Consider the case in which the antigens are part of invading disease organisms: when anti-

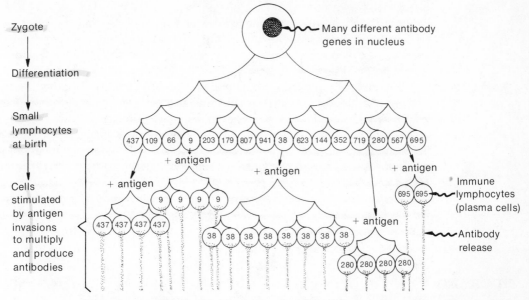

Figure 11–2 Clonal selection theory. Numbers refer to cell-specific antigen sensitivity and also to the various antibody-determining genes expressed in different lines following stimulation. (Modified from Edelman, G. M.: Antibody structure and molecular immunology. Science, *180*:830–840, 1973.)

bodies are complexed to these antigens the invading cells either will break up (lyse), owing to the activation of certain enzymes (collectively called complement), or will clump together (agglutinate). Individuals are said to be immune to a disease when their body contains specific antibodies which stick to the invading cells and prevent them from proliferating and causing symptoms of illness.

HOW DOES IMMUNITY COME ABOUT?

The fetus receives antibodies from its mother through the placenta and the infant, if fortunate enough to be breast fed, continues to receive them from breast milk. These antibodies do not last long, but upon exposure to various microorganisms the infant starts producing antibodies against them. Such acquired immunity may last for years, even for a whole lifetime for certain kinds of diseases. In other cases, such as for smallpox, immunity lasts only three years or so.

Vaccination is a technique which exploits the phenomenon of acquired immunity, but does so without the necessity of a "real" infection. In contrast to an accidental infection by live virulent organisms, in vaccination one is deliberately infected by either killed microorganisms or live non-virulent relatives of the virulent strain. In either case the foreign entities persist long enough in the vaccinated individual for his body to be "tricked" into producing the antibodies which by human ingenuity happen also to be antibodies against "the real thing." This is the explanation for the seminal observation by Edward Jenner in 1798 that milkmaids exposed to the pox disease of cows (cowpox) had become immune to the pox disease of man (smallpox). (The Chinese knew over 2000 years ago, however, that inhaling dried crusts from the sores of people with smallpox produced immunity to the disease.)

A few children are unable to produce antibodies of any kind. Since antibodies are found among the so-called gamma globulins in the blood serum, such children are called agammaglobulinemics. Until recently they invariably died soon after birth from common kinds of infections. Now a few are kept alive, at great cost, by the surgical transplant of bone marrow cells from close relatives who do not have the condition. Several such children are even being raised in a germ-free environment in their own self-contained plastic bubble. The defect of agammaglobulinemia is an autosomal recessive trait.

In contrast to agammaglobulinemics, normal individuals produce a very large number of antibodies during their lifetime. The body is stimulated to produce antibodies as a result of invading antigen molecules binding to receptor proteins on the surface of small lymphocytes (Fig. 11–2). Lymphocytes are white cells produced in the thymus, spleen, and bone marrow and are located predominantly in those areas and in the lymph nodes. The antigen-bound lymphocytes become larger and undergo mitosis; their cell-line descendants (a clone of plasma cells) start producing antibody specifically against the invading antigen, thereby bringing about its destruction. Each cell in the clone can produce and secrete several thousand identical antibody molecules per second.

BLOOD GROUPS

The Rh System. A genetically simple system of blood incompatibility with peculiar medical effects is the Rh immunity system. If you are Rh-negative (Rh$^-$) and receive a blood transfusion from an individual who is Rh-positive (Rh$^+$), your body will undergo an immune response. Certain molecular (antigenic) parts of the Rh$^+$ red blood cells cause this chemical-cellular reaction when those cells are introduced into an Rh$^-$ person. The lymphocytes of the Rh$^-$ recipient start producing antibodies which circulate in the blood, bind to the surface of the invading Rh$^+$ red cells, and destroy them. Because these antibodies specifically attack the Rh$^+$ characteristics of the invading cells, they are called anti-Rh$^+$ antibodies. Since one antibody molecule can react with two cells at once, and since each cell has many antigenic sites, this antigen-antibody reaction causes the foreign cells to clump together, that is, to agglutinate. In the case of an Rh-incompatible blood transfusion, the clumping produces devastating effects in the recipient: fever and symptoms of shock (complicating whatever condition the transfusion was needed for in the first place). When transfusions first became popular in medical practice in the nineteenth century, some patients died because their blood was Rh-incompatible with that of the blood donor. Nowadays, however, individuals are tested for blood type and transfusions are performed only between people who are Rh-compatible (and ABO-compatible, as discussed later).

To determine an individual's Rh status, a drop of his or her blood is mixed in the laboratory with known anti-Rh$^+$ antibodies. If the red cells clump together in the presence of these antibodies, the individual is Rh$^+$; if they do not clump, he or she is Rh$^-$ (Fig. 11–3). Because Rh$^-$ individuals have *no* Rh antigens on the surface of their red cells, Rh$^-$-type blood can safely be used for transfusion if other blood factors are compatible. Rh$^-$ blood will not elicit an antibody response in the recipient.

Rh$^+$ is a dominant phenotype (genotypes DD and Dd) and Rh$^-$ is recessive (dd). Various investigators have worked out fine gradations among the Rh phenotypes that probably reflect minor differences in a variety of mutant alleles, but these distinctions are not important for our purposes. Mankind is polymorphic (genetically variable) for Rh alleles throughout the world, although the frequency of Rh$^-$ is much lower in Asia than it is in Europe. In China and Japan, Rh$^-$ is very rare indeed.

Phenotype	Antigens (▵) * existing on red cells	Rh-type antibodies which can be produced	Results of test of Rh phenotype by mixing a drop of blood with known anti-Rh⁺ antibodies (X)
Rh⁺	Rh⁺	None	Clumping
Rh⁻	None	Anti-Rh⁺	No reaction

Figure 11–3 Rh blood group system.

*greatly magnified

In the U.S. and Europe the frequency of Rh⁻ is high enough that medical complication of the Rh genetic system occurs fairly commonly. Among whites in the U.S., one out of every eight marriages is between an Rh⁻ woman and an Rh⁺ man. If such a woman conceives an Rh⁺ child, mother-fetal incompatibility for Rh blood type results. Since fetal blood cells rarely penetrate into the mother's circulation through the placental barrier, this mother-fetal incompatibility produces no adverse effects on most such Rh⁺ offspring. But occasionally the pregnant Rh⁻ mother accidentally receives a small "transfusion" of Rh⁺ fetal blood through the placenta and her immune system starts producing anti-Rh⁺ antibodies. Since these antibody molecules are proteins which *can* diffuse through the placenta into the fetal bloodstream (just as other molecules pass through, to nourish the fetus), the mother's induced anti-Rh⁺ antibodies can enter the fetus and agglutinate its Rh⁺ red cells, producing an anemia. If this happens the fetus may be stillborn or born with a deficit of mature red cells and an excess of immature red cells, the erythroblasts. The infant, who will appear jaundiced at birth, is said to have erythroblastosis fetalis. A newborn "Rh baby" is given massive blood transfusions to allow its system to rebuild its population of mature red cells. After a week or so the infant is usually out of danger. Once born it can no longer be infiltrated by its mother's antibodies.

The discussion so far has outlined the worst possible outcome of mother-fetal Rh incompatibility. For many years now the medical profession has known of this problem and pregnant women are routinely blood-typed for Rh when they seek obstetrical care. Suppose a woman is found to be Rh⁻ on her first pregnancy. If her husband is also Rh⁻, there is no problem. But if he is Rh⁺, she will be checked periodically during the pregnancy to determine whether she is developing anti-Rh⁺ antibodies. If she does experience an antibody build-up, it will typically occur near the end of pregnancy, because earlier the fetus is too small to make enough red cells to cause a problem. Birth labor can be induced prematurely to protect the fetus from substantial exposure to the mother's induced antibodies.

Usually, however, only a negligible antibody build-up occurs during a first, or even a second, pregnancy. With successive pregnancies the chance of an antibody build-up increases in general. But the chance of such a build-up increases particularly because of the birth process itself, since during childbirth there is the greatest danger of mixing of maternal and fetal blood. When the placenta is detached from the uterine wall in labor, a small amount of fetal blood in the placenta usually leaks into the mother's bloodstream. The placenta (expelled in the afterbirth) is, of course, the site where oxygen, food, and other materials are supplied to the fetus by the mother through the close juxtaposition of the maternal and fetal blood vessels.

Rh⁻ women who give birth to an Rh⁺ child are now routinely given an injection of a small amount of anti-Rh⁺ antibodies immediately following delivery. The antibodies complex with and destroy any fetal Rh⁺ cells in the mother's blood before her immune system itself has had the chance to trigger the production of such antibodies. The injected antibodies do not long persist in the mother's body. She is made virtually "innocent" immunologically by the treatment and need not be apprehensive, on this account at least, about a subsequent pregnancy.

The ABO System. Let us now turn to the best known blood group system: ABO. Here, in contrast to the Rh system, a key point is the multiple allelic situation. In addition, antibodies for ABO are present long before birth; their production is probably stimulated by the presence of harmless bacteria in the gut. Like the Rh system, ABO is a worldwide polymorphism, again with differences among distinct populations (Table 11–1). Also, ABO is important in determining transfusion compatibility, and, to a lesser extent, mother-fetal incompatibility.

ABO is the system in which genetic differences in blood were first recognized. In 1900 Landsteiner mixed together a few drops of blood from pairs of different persons. Some mixtures showed clumping and other mixtures did not, thereby providing an explanation of some of the transfusion failures of earlier years. Landsteiner designated his subjects as being blood type A, B, O, or AB. Subsequent research showed that three major alleles determine this system, forming the four phenotypes. Blood group A is genotypically homozygous (AA) or heterozygous with the O allele (AO). Blood group B is BB or BO in genotype. Allele O is thus recessive to both A and B. Blood group AB is designated AB genotypically also. A and B are thus codominant with each other. Blood group O is genotypically OO. (This is the simplest description of the system.)

Individuals with blood group A have A-type antigens on their red cells and have anti-B antibodies in their blood serum. Clearly, these individuals are transfusion incompatible with blood donors who have B antigens on their red cells (that is, donors with blood types B and AB). The other antigen-antibody relationships in this system are shown in Table 11–1 and Figure 11–4.

Offspring which are incompatible in utero with the mother's blood type are produced from certain kinds of ABO matings. For example, an AO mother and a

TABLE 11–1 ABO Blood Groups*

Phenotype	Genotype	Antigen(s) on Red Blood Cells	Antibodies in Blood Serum	Frequency in Some Populations		
				W. African Black	U.S. Black	U.S. White
A	AA or AO	A	anti-B	.23	.25	.41
B	BB or BO	B	anti-A	.24	.20	.07
O	OO	none	anti-A & anti-B	.50	.51	.50
AB	AB	A and B	none	.04	.04	.02

*Population data from P. L. Workman, B. S. Blumber, and A. J. Cooper. Selection, gene migration, and polymorphic stability in a U.S. White and Negro population. American Journal of Human Genetics 15:429–437, 1963.

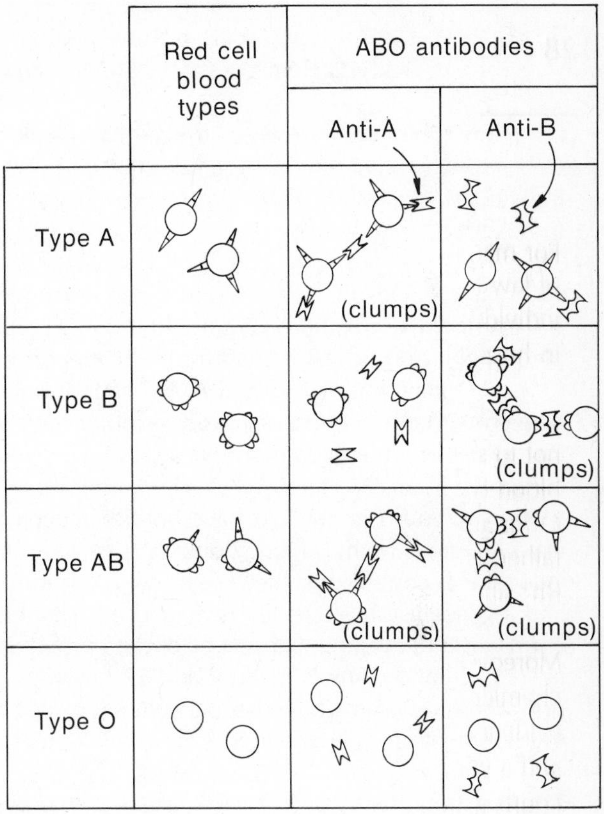

Red cell blood types	ABO antibodies	
	Anti-A	Anti-B
Type A	(clumps)	(clumps)
Type B		(clumps)
Type AB	(clumps)	(clumps)
Type O		

Figure 11–4 ABO blood group system. Red blood cells of various types are mixed with known antibodies to determine phenotype.

BO father can produce a BO offspring (25% chance). But since the anti-B antibodies of the AO mother pass through the placenta only with difficulty, their reaction with the B-type fetal cells is characteristically weak, and only a mild erythroblastosis (anemia) results. Still, this occurs in one in about 1000 pregnancies among U.S. whites, making it twice as common as Rh disease of the newborn. On the other hand, if some B-type fetal red cells pass into the mother's circulation, her anti-B antibodies will quickly destroy them.

Curiously, it appears that such ABO incompatibility between mother and fetus is advantageous to the fetus if the parents are also Rh incompatible. In the example above, if the AO mother is also Rh-negative (dd) and the BO father is Rh-positive (DD or Dd), their BO offspring will have both Rh$^+$ and B antigens on its red cells. Since the mother's anti-B antibodies destroy such fetal cells in her bloodstream upon contact, there is no triggering of production of anti-Rh$^+$ antibodies in her body. The fetus in such circumstances will probably escape being severely anemic at birth.

Other Systems. Many other blood-type genetic systems are known. They are controlled by diverse genes scattered throughout the chromosomes. For most of them, human populations have few variants, that is, the populations are almost monomorphic (most individuals being homozygous for the common allele). One of the significantly polymorphic blood group systems not yet mentioned is MN. Here there is no difficulty with blood transfusions, as the antibodies involved are presumably weak ones. There are three phenotypes (and genotypes): MM, MN, and NN. As in the case of A vs. B, M and N are codominant alleles.

FATHERHOOD: EXCLUSION AND INCLUSION

For many years blood group data have been used as evidence in courts of law to resolve questions of fatherhood in paternity suits, to identify individuals (or bodies), and to correct accidental mixups between infants in hospitals.

The courts in the U.S. allow blood group evidence only to exclude paternity, not to show the possibility of paternity. For example, suppose a man with blood type Rh^+ and AB is sued for child support by a mother with Rh^- and O for an infant who is Rh^+ and A. He cannot be excluded as the possible father. Neither, of course, can a lot of other men be excluded, since both Rh^+ and A are quite common throughout the country.

Moreover, courts in the U.S. have not yet accepted polymorphic biochemical traits such as hemoglobin and various enzyme defects as routine evidence in cases of disputed paternity. Margery Shaw, who is both a lawyer and a geneticist, has recently reviewed this situation and contends that the courts are not availing themselves of the current wealth of knowledge about genetic variation. She points out that 57 blood group and biochemical polymorphic systems now can be routinely tested for such purposes as the development of legal evidence.

There is, however, at least for the testing of blood groups, at least a two per cent misclassification of individuals. The extent of misclassification for biochemical traits is unknown, but is thought to be less than two per cent. Still, this amount of error in classification is a real drawback in allaying reasonable doubts in law. But if a putative father were tested for all 57 genetic systems now available, the chance of his being wrongly accused of paternity on the basis of genetic data is very low.

A generation ago, paternity suits in which genetic data were used as evidence were often headline cases of a rich man being sued by a rather poorer woman. Even Charlie Chaplin was once sued in such circumstances. Though such cases still occur they are rarer now, probably because of the greater economic independence of women today (not to mention the availability of means to avoid or arrest an unwanted pregnancy). Indeed, medical geneticists sometimes come across test results on a family in which a blood trait (or biochemical trait) of a child is not in agreement with that of its "father" — but rarely, if ever, do they mention this discrepancy to the woman or her husband, since it might disrupt the family as a functioning unit.

TRANSPLANTATION

From studies on humans and on the laboratory mouse, it is now understood that a complex genetic situation is involved in the common rejection of transplanted hearts, kidneys, skin, and other organs from one person to another. Here again we see a lavish diversity in genetic makeup among individuals. Even our tissues are genetically so distinctive they are intolerant of tissues from other human beings. (They are, of course, even more intolerant of tissues from other species.)

A major reason for tissue incompatibility and thus graft rejection in human beings is genetic diversity in the HL-A system. HL-A, which stands for human leucocyte-A, has two gene loci, LA and Four, each having over 15 alleles. Thus the chance of any two unrelated individuals possessing the same alleles at both loci is vanishingly low (far lower than one in 10,000, it has been estimated). A tissue graft is almost always rejected by an individual if the tissue donor has any HL-A alleles that are different from those of the tissue recipient.

Tissue transplants are rejected by the human body as a result of the immune system's recognition of their alien molecular composition. Unlike the situation with blood transfusions, however, the immune response to a heart transplant does not involve the production of antibodies. Instead, the so-called T lymphocytes (which have matured in the thymus gland) multiply in response to antigens on the surface of the alien cells and attack the surgically implanted tissue (Fig. 11–5). Moreover, the attacking T lymphocytes secrete chemical substances which attract cells called macrophages to the site of invasion. The macrophages digest the foreign cell parts.

To suppress rejection of a tissue transplant, immunosuppressive drugs like cortisone are administered to the patient. This procedure is sometimes fatal in itself, however, as the body is then bereft of means to combat ordinary infections.

AUTOIMMUNITY

Why doesn't the body become immune to its own cell antigens? The body distinguishes foreign disease organisms and other foreign antigens (non-self) from its

Figure 11–5 Two aspects of the immune response.

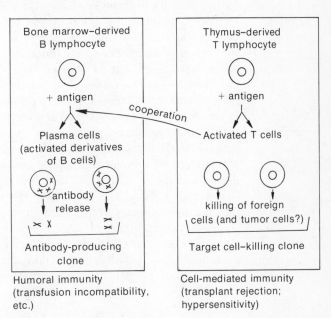

Bone marrow–derived
B lymphocyte

+ antigen

cooperation

Plasma cells
(activated derivatives
of B cells)

antibody
release

Antibody-producing
clone

Humoral immunity
(transfusion incompatibility,
etc.)

Thymus–derived
T lymphocyte

+ antigen

Activated T cells

killing of foreign
cells (and tumor cells?)

Target cell–killing clone

Cell-mediated immunity
(transplant rejection;
hypersensitivity)

own cells (self). This statement, however, does not hold for identical twins or inbred strains of mice—here brothers and sisters do not recognize each other as being different in terms of their body's immune response.

Moreover, some individuals do become immune to their own antigens. Such autoimmunity results in the immune system damaging a person's own tissues. The body is then no longer self-tolerant immunologically. Autoimmunity probably underlies diseases like rheumatic fever, certain anemias, and systemic lupus erythematosus. Rheumatic fever usually follows a "strep throat" or other streptococcal bacterial infection. The bacteria evidently trigger the production of antibodies against the individual's own heart tissue, a reaction which weakens the heart and may even cause heart failure. The bacteria and heart tissue happen to have very similar antigenic determinants on their surfaces; this is the connecting link between the two diseases and the cause of the mistake in the response by the immune system.

In systemic lupus, another disease in which autoimmunity is implicated, genetic factors probably play a role in the susceptibility to the production of lupus antibodies. The evidence for this, however, is equivocal. The disease affects about half a million people, most of whom are women. It usually develops between the ages of 20 and 40 years and varies between mild and severe in terms of damage to internal tissues, particularly the kidneys. For reasons which are not clear, people with lupus have antibodies against nucleic acids (usually DNA); it is the deposition of these antibodies with DNA antigens in the kidneys that causes illness. These antibodies may not be strictly against "self" DNA, but may possibly be against the gene material of parasitic viruses. Here we see that genes, viruses, and perhaps other as yet undiscovered factors may all interact in certain disease problems.

AGING

Is aging just an extension of immunogenetic problems? The thymus gland, which is located behind the breastbone, is at its largest in the newborn and then declines with age. Many functions of the immune system also decline with age. Since the thymus is necessary for T lymphocyte functions in immunity, one might ask whether these changes in the immune system are the cause of the increase in infections, autoimmune disease, and cancer that occurs with advanced age. Is the thymus gland, as MacFarlane Burnet suggests, the pre-set biological clock that runs down as we age? Are there longevity genes in the thymus, or elsewhere, which are pre-set ? (See Table 11–2.)

Some research workers favor immunological explanations for aging processes. Others favor an explanation based on the idea that cells are genetically programmed to run down (and stop dividing) at a predetermined time. Still others think random genetic damage (mutations) such as chromosomal alterations or mutations in inactive (non-RNA producing) DNA may explain aging. To be sure, mutations generally hinder cell survival and mitotic ability. But the validity of explanations is not determined by the speed with which they are adopted or by the number of those who embrace them. The problem in research on aging is obviously not a lack of theories; the field may in fact suffer from an oversupply of them.

Although no treatment has yet been developed to lengthen the human life span greatly beyond the eighth or ninth decade, very old invididuals do exist in many populations, particularly where a non-affluent (low-calorie) but balanced diet and physical labor prevail. In laboratory mice, experimenters have also found that a

TABLE 11-2   The Longevity of Offspring as Affected by the Longevity of Parents*

Father's Life Span (years)	Mother's Life Span (years)		
	Under 60	60–80	Over 80
	Offspring life span (years)		
Under 60	32.8	33.4	36.8
60–80	35.8	38.0	45.0
Over 80	42.3	45.5	52.7

*Data based on a study by Alexander Graham Bell of over 2200 male and 1800 female descendants of one William Hyde. (From Bell, "The duration of life and conditions associated with longevity. A study of the Hyde geneology." 1918. Washington, D. C.)

non-affluent diet produces much longer-lived mice (e.g., a 50% increase in life span). Such a diet also slows the maturation of the immune system in young mice; but as the mice age their immune system is more active than that of mice on a conventional diet. However, different inbred strains of mice (different genotypes) also differ substantially in their life span.

The experimental "votes" are not all in yet on the question of the causes of aging. But, in considering our present lack of immortality, it may be helpful to keep in mind the following:

> Which is better, to be young or to be old? To be old is to have less time before you and more mistakes behind you. I leave you to decide whether this is better or the reverse*

*Idries Shah: Thinkers of the East. Baltimore: Penguin Books, 1970, p. 40.

Old age among the people of Malaysia.

SUMMARY

Since the nineteenth century, when it was first realized that people were blood incompatible for transfusions, the field of immunogenetics has become important for understanding a host of disease states in human beings. Immunological individuality is much more pronounced than, say, metabolic individuality in terms of enzyme variation. This has advantages and disadvantages. It is advantageous for the body to recognize itself as being different from foreign invaders and for it to defend itself against their invasions. The disadvantage is in the body's intolerance of certain kinds of blood transfusions and tissue transplants. But, overall, this mixed blessing of immunological response is more useful than not. Its role in the complex interactions of aging is currently an issue of hot debate.

STUDY QUESTIONS

1. The MN and Rh genes are unlinked. What is the chance of dd MN offspring from a mating of Dd MN × Dd NN as a result of independent gene segregation?

    (a) $\frac{1}{2}$ (b) $\frac{1}{4}$ (c) $\frac{1}{8}$ (d) $\frac{1}{16}$

2. The MN and ABO genes are unlinked. What is the chance of an AB MN offspring from a mating of AO MN × BO MN as a result of independent gene segregation?

    (a) $\frac{1}{2}$ (b) $\frac{1}{4}$ (c) $\frac{1}{8}$ (d) $\frac{1}{16}$

3. If you did a family survey, which kind of maternal ABO phenotype would you expect to show the greatest incidence of newborn anemia and other indications of mother-fetal incompatibility?

    (a) A   (b) B   (c) O   (d) AB

4. Surveys of marriages of AB × OO indicate an offspring ratio very close to $\frac{1}{2}$A : $\frac{1}{2}$B. This ratio is often cited as part of the evidence for A, B, and O being alleles of the same gene. What offspring types would be expected if A and B were not allelic and O was simply the antigen-deficiency allele at both gene sites, that is, if these marriages were AA BB × OO OO genotypically?

5. In the chapter it was stated that internal defense to disease was chiefly mediated by the immune system. What other kind of internal defense against disease do you know of?

6. Key terms:

antibody	Rh	HL-A
antigen	ABO	monomorphic
immune reaction	incompatibility	autoimmunity
blood type	polymorphic	

SUGGESTED READINGS

Beer, A. E., and R. E. Billingham. The embryo as a transplant. Scientific American 230: 36–46, 1974.

Benacerraf, B., and H. O. McDevitt. Histocompatibility-linked immune response genes. Science 175:273–279, 1972.

Buche, K. D., and R. C. Elston. Estimation of nonpaternity for X-linked trait. American Journal of Human Genetics 27:689–690, 1975.

Burnet, M. Intrinsic Mutagenesis. A Genetic Approach to Aging. New York: John Wiley & Sons, Inc., 1974.

Chakraborty, R., M. Shaw, and W. J. Schull. Exclusion of paternity: the current state of the art. American Journal of Human Genetics 26:477–488, 1974.

Chandra, H. S. Mother-child incompatibilities for ABO and Rh alleles: possible association with certain types of chromosomal aberrations. New England Journal of Medicine 272:566–569, 1965.

Clarke, C. A. The prevention of "Rhesus" babies. Scientific American 219:46–52, 1968.

Comfort, A. Aged Equadoreans. Nature 258:41, 1975.

Edelman, G. M. Antibody structure and molecular immunology. Science 180:830–840, 1973.

Humphrey, J. H. Some implications of modern immunology. In The Biological Revolution. W. Fuller (ed.). Garden City, New York: Doubleday & Co., Inc., 1972, pp. 147–168.

Jerne, N. K. The immune system. Scientific American 229:52–60, 1973.

Marx, J. Aging research: pacemakers for aging? Science 186:1196–1197, 1974.

Marx, J. Thymic hormones: inducers of T cell maturation. Science 187:1183–1185 and 1217, 1975.

Marx, J. Antibody structure: now in three dimensions. Science 189:1075–1077, 1975.

Notkins, A. L., and H. Koprowski. How the immune response to a virus can cause disease. Scientific American 228:22–31, 1973.
Some virus diseases, such as hepatitis, are caused not by the virus, but by the lymphocytes attacking the virus.

Osher, F. C., and W. C. Neal. Theories of antibody diversity: the great debate. Cellular Immunology 17:552–559, 1975.

Reisfeld, R. A., and B. D. Kahan. Markers of biological individuality. Scientific American 226:28–37, 1972.

Strehler, B. L. The understanding and control of the aging process. In Challenging Biological Problems. J. A. Behnke (ed.). New York: Oxford University Press, 1972, pp. 133–147.

Yielding, K. L. A model for aging based on differential repair of somatic mutational damage. Perspectives in Biology and Medicine 17:201–208, 1974.

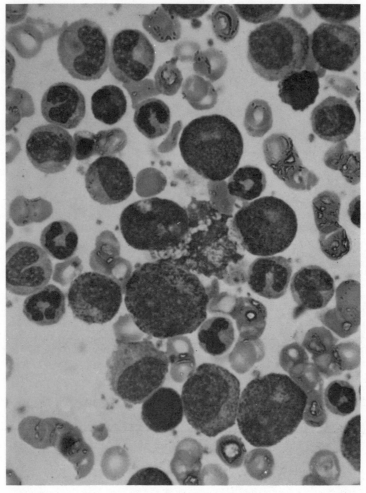

Acute myelogenous leukemia, 100 ×. (From Polaroid Corporation, Cambridge, Mass.)

It would be nice to believe that we now possess all the basic scientific facts needed for the conquest of cancer. We could then set up timetables of expected progress, hire the required hands, and confidently expect this awful scourge to disappear from the human vocabulary. The truth, however, does not fit this fantasy and despite the stunning achievements of the modern biological revolution that has revealed the nature of the gene as well as the operation of the genetic code, huge gaps still exist in our knowledge of the fundamental biology and genetics of normal human cells, much less of their cancerous counterparts.

J. D. Watson

Chapter Twelve

VIRUSES AND CANCER

Is a cancer cell a genetic system which has gone wild or has been disabled? Cancer is the overproduction of cells in the body. In this chapter we shall inquire if cancer is due to a virus attack—a popular belief in some quarters—and, in particular, we shall try to unravel the genetic clues to cancer formation.

VIRUSES

A great deal has become known about viruses since the nineteenth century, when their existence was first suspected. These minute bodies, some no bigger than a chromosome of three or four genes surrounded by a protein coat, live as parasites in many plants, animals, and bacteria. As viruses are virtually only nomadic chromosomes, they are smaller than any known cell but larger than molecules. They can reproduce only inside a host organism, subverting its food and energy supplies, its enzymes, and its ribosomes for the production of descendent viruses. In a world full of ingenious genetic systems, viruses are uniquely streamlined for the task of reproduction.

Various plant RNA viruses cause leaves to wilt, become splotchy, or develop lumps. These viruses are often a problem to plant breeders and farmers because they can cause disastrous epidemics in food crops. Bacterial viruses, known as bacteriophages (or phages), attack bacteria that live in the soil, water, human gut, and elsewhere. Human viruses are quite diverse. Some, but not all, cause disease: smallpox, flu, polio, mumps, the common cold—to mention only a few. Human viruses are transmitted from person to person in water or food, or through the air. Viral epidemics were common in the past, and could occur again. The most devastating epidemic known was the 1918 flu epidemic, which killed an estimated 20 million people worldwide. Smaller flu epidemics have occurred recently, in 1957 and 1968. New strains of flu virus with new hereditary capabilities gain ascendancy in the world as the target population builds up immunity to older strains. Each epidemic is associated with the flare-up of a new genetic type.

Some viruses contain DNA genes. Others, no less interesting, contain RNA as their genetic material. Polio virus, for example, is a small spherical virus with a single strand of RNA about 600 bases long in the viral core. The core is enveloped by a protective layer of globular protein molecules. Larger viruses also contain a few enzymes, along with their nucleic acid packaged inside their protein coat.

Viruses have a life cycle, as does the rest of the living world. Infection begins when a virus bumps into a cell that is susceptible to its attack and attaches to the cell at a specific membrane site. A cell is susceptible only if it has the correct chemical conformation on its surface for viral attachment (just as certain cells have specific antigen on their surface, as discussed in Chap. Eleven). The attached virus then squeezes out its nucleic acid through the cell membrane into the cytoplasm of the host cell. The viral coat, which protected the nucleic acid core in its wanderings, usually remains outside the cell as a useless vestige.

Some viruses have only a lytic type of life cycle. This means that inside the host cell they multiply and then lyse (i.e., break open) the cell when its resources are used up. In such cases, once the viral nucleic acid is inside the cell it begins to take over the cell's machinery for the business of making more viruses. For a human RNA virus like polio virus, this means that the viral RNA acts as mRNA at the ribosomes for the production of viral proteins (Fig. 12–1). The viral RNA replicates by Watson-Crick pairing rules, producing a complementary strand. This complementary strand then makes many strands complementary to it (but identical to the infecting strand of RNA). These final RNA strands combine with the viral proteins made earlier to constitute mature viruses, which spill out of the cell upon lysis, ready to infect other cells with which they come in contact.

The lytic cycle of a human DNA virus like adenovirus, which causes respiratory infections, is similar to that outlined above. The linear, double-stranded DNA molecule of adenovirus replicates in ordinary, semi-conservative fashion. But in

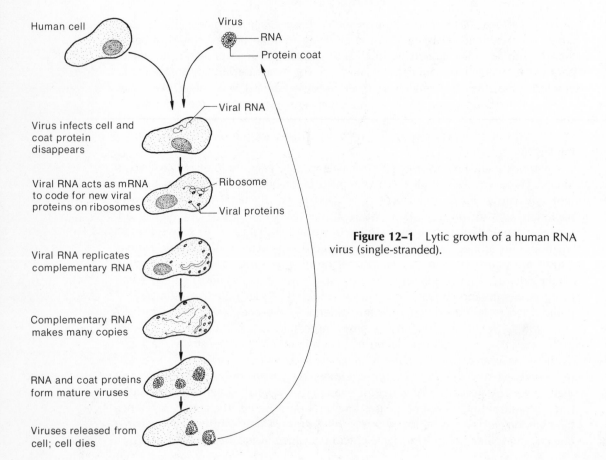

Figure 12–1 Lytic growth of a human RNA virus (single-stranded).

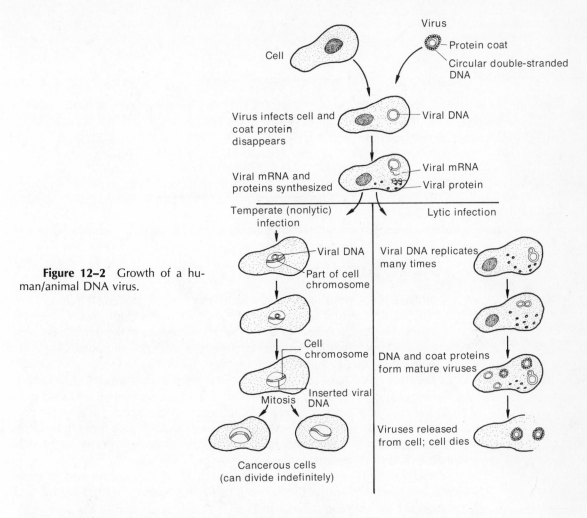

Figure 12–2 Growth of a human/animal DNA virus.

the case of the DNA virus called SV40, which multiplies well in monkeys but apparently not in humans, there is the interesting complication that the viral DNA is a double-stranded ring. The ring DNA replicates more double-stranded rings, again by semi-conservative mechanisms, and produces mRNA which is decoded into viral proteins on the cell's ribosomes. After assembly of mature viruses, the cell lyses and releases the viruses. If genotypically different viruses of one "species" infect the same cell, they can undergo genetic recombination by a process akin to crossing over in cell nuclei and thus produce progeny viruses with new combinations of genes.

Some viruses, some of the time, are capable of living in a "temperate" state inside the host cell. Instead of multiplying independently of the host's genetic material, and eventually lysing the cell, the viral DNA can be inserted into the host DNA and can replicate as an integral part of it. For example, SV40 virus behaves temperately when experimentally introduced into hamster cells in culture, but it can only exist lytically in green monkey cells. In the temperate state, SV40 causes some of the hamster cells to change (transform) into tumor cells (Fig. 12–2).

A certain class of single-stranded RNA viruses, some of which cause tumors in one or another test animal, are considered prime candidates for causing certain cancers in humans (despite the fact that proof does not exist because humans can't be experimentally infected with such viruses). These so-called RNA-DNA

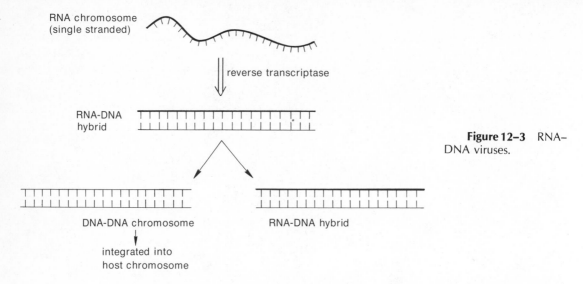

RNA chromosome
(single stranded)

reverse transcriptase

RNA-DNA
hybrid

Figure 12–3   RNA–DNA viruses.

DNA-DNA chromosome   RNA-DNA hybrid

integrated into
host chromosome

viruses all have the curious property of transcribing single-stranded DNA from their single-stranded RNA template, using an RNA-directed DNA polymerase (reverse transcriptase). The single strand of DNA then replicates to make a Watson-Crick DNA molecule which is then integrated into the host cell's chromosomal DNA (Fig. 12–3). With a so-called nonproductive infection, the integrated viral DNA (called a DNA provirus) either has no visible effect or it transforms the host cell into a cancer cell. Subsequently the provirus transcribes viral RNA for translation into viral proteins in the case of a lytic, or productive, infection. (More about this after we discuss cancer in general terms.) H. M. Temin has postulated that these RNA-DNA viruses arose as normal cell genes which have become independent entities through mutation or abnormal genetic recombination.

Several practical problems that have arisen in virus vaccines and in virus research are discussed in the Boxes on Virus-Contaminated Vaccines below, and Experimental Hybridization of DNA (p. 139).

VIRUS-CONTAMINATED VACCINES

Vaccination against disease has saved countless lives. For diseases like polio, measles, and mumps the current vaccines contain live but harmless viruses that are antigenically similar to their disease-causing relatives. These vaccines are usually prepared by culturing virulent viruses in non-human cells (such as monkey cells). As the virulent virus adapts to the culture conditions by genetic changes (mutations), it loses its virulence for human cells. Indeed, it is often a problem in preparing vaccines to keep the mutant virus virulent enough so it will be able to get into (infect) human cells later. When the attenuated (i.e., weakened) virus is administered to people as a vaccine, the human body builds up antibodies against it, thereby becoming immune to subsequent infection by the virulent relative.

So far so good. But it turns out that a viral contaminant existed in one of the live-virus vaccines in the past. The contaminant is the virus SV40, which grows slowly and harmlessly in cultured cells of rhesus monkey but causes transformation in cultures of hamster or human cells. (Transformed cells frequently cause tumors when inoculated into an animal.) SV40 got into polio vaccine because the polio virus was cultured in rhesus monkey cells which already harbored SV40. There is no indication to date that this accidental contamination by SV40 has increased the incidence of human cancer or affected human health in any way (but see J. J. Holland's article (1974), cited in the Suggested Readings at the end of this chapter). Since the recognition of this contamination, however, techniques have been changed to ensure that SV40 is now excluded from batches of polio vaccine during the production process.

The SV40 story unfolded in the early 1960's. Another type of contaminant of vaccines came to light in 1972: a government agency announced that all live-virus vaccines (for mumps, polio, etc.) are contaminated by bacterial viruses (phages). This occurs because blood serum from fetal calves is used to provide nutrients for the cell cultures in which the attenuated viruses grow. The fetal calf serum is collected from slaughterhouses which are infested with enormous populations of bacteria. The collected serum is filtered to remove bacteria and debris, but the phages of the bacteria pass through the filters (as do many viruses, because they are so small). The phage-containing serum is then used directly in the cell cultures to nurture the vaccine-type virus. The government now requires that serum be sterilized before it is used in vaccine production.

The worry about phage contamination of vaccines is at least twofold. First, diseases like diphtheria are called bacterial diseases although they are actually caused by a toxic substance produced by phages growing in the bacteria. A person with phage-free diphtheria organisms living in his gut could perhaps contract the disease by taking a live-virus vaccine contaminated with diphtheria-type phage. Secondly, certain kinds of phages have now been found to reproduce in cultured hamster cells. Might they also be able to adapt to living in human cells? Most investigators doubt it, but then no one yet knows for sure.

EXPERIMENTAL HYBRIDIZATION OF DNA

Recently several enzymes have been discovered and isolated that can snip DNA molecules into gene-size lengths. The pieces of DNA, which have "sticky" ends, can then be joined to the DNA of different organisms. With the aid of these "restriction endonuclease" enzymes, some foreign genes have been experimentally inserted into bacterial cells in the laboratory. This new technique might prove dangerous to humanity if the artifically hybridized cells escaped into the outside world. For example, the fast-multiplying bacterium *Escherichia coli* is a favorite laboratory organism for genetic studies, but it is also (along with other bacteria) a harmless resident of the human gut.

Some disease-causing bacteria that can invade the gut can recombine genetically with *E. coli* by natural processes. If foreign genes which confer resistance to antibiotic drugs or confer the ability to produce toxic substances are integrated into the DNA of *E. coli* cells, there is at least the potential for harm to humans, either by mischance or mischief. Moreover, the genes of animal viruses, including tumor viruses, can also be experimentally integrated into the DNA of bacterial cells. Bacterial cells with such viral-bacterial hybrid DNA may conceivably express new, bizarre phenotypes which might endanger human life. Finally, the insertion of genes (using frog and insect DNA) from animal cells into bacteria has already been accomplished. This raises the possibility of accidentally transferring from animal cells to bacteria genes which have the capacity to trigger multiplication of tumor viruses in those bacteria. No one knows if this *would* happen — but the loss of such tumor-causing bacteria down the laboratory sink is not a welcome prospect. As Erwin Chargaff has said, "You can stop visiting the moon . . . you may even decide not to kill entire populations by the use of a few bombs. But you cannot recall a new form a life."*

Because of these considerations, in 1975 an international group of researchers headed by Paul Berg called for stringent safeguards on such DNA manipulation experiments. One safeguard they advocated was the construction of bacterial "guinea pigs" that would self-destruct if they ever escaped from the laboratory. That is, mutant strains would be used in the laboratory that would be incapable of survival under normal conditions of food supply, temperature, and so on. Since 1975, agencies of several national governments have drawn up restraining regulations about such DNA manipulation experiments. In effect, however, proponents of DNA manipulation have defined the problem and proposed their own solutions without informed public consent.

In contrast to this somber perspective, some scientists point out the potential use of artificially hybridized DNA for molecular engineering of genetic improvements (genetic engineering). One small but intriguing possibility is the manufacture of large quantities of insulin by bacteria in the laboratory. This would require isolation of the correct human gene, its insertion into bacteria, and the growth of vat-sized batches of the reprogrammed bacteria for a large insulin harvest. Such a cheap supply of insulin for treating diabetes would be technologically attractive. The same can be said for adopting this method to mass produce an abundance of specific antibodies for the treatment of other human diseases.

Thus we have, on the one hand, moral censure by a conscientious group of scientists against anyone doing certain kinds of experiments, or at least against doing them without extreme safeguards. On the other hand, there is human curiousity about a kind of genetic manipulation that could lead to advances both in understanding how and why genes act and in developing sophisticated gene machines for producing "crops" of useful gene

---

*E. Chargaff. On the dangers of genetic meddling. Science *192*:938–940, 1976. Quote is from p. 938.

products. The experimental risks, though conceivable and dangerous, seem far-fetched; the benefits are tempting. Which consideration should outweigh the other?

There is also a political-legislative problem here, as R. Sinsheimer pointed out in 1975. Discussing comments of J. D. Watson and others on this problem, Sinsheimer states: "Watson says quite correctly that there is no way to measure the risk. But it seems to me that in the end we [scientists] will be regulated [by legislation]. We would be in a better position to face that if we take the position that some of the higher [risk] experiments should not be done until more information is available. I can't think of anything that would impede science more than an epidemic [caused by artifically hybrid DNA]. . . ."*

We can only hope that long-range, prudent views prevail among the research scientists interested in such experiments.

---

*Quoted by N. Wade: Genetics: conference sets strict controls to replace moratorium. Science *187*:931–935, 1975.

CANCER

Cancer is a complex of many diseases. Over 100 distinct types are recognized. There are four main categories: carcinomas, sarcomas, lymphomas, and leukemias. Eighty-five per cent of the cancers are carcinomas, the solid tumors in nerve tissue and in tissues of body surfaces or their attached glands. Thus there are cervical, breast, skin, and brain carcinomas. Sarcomas are solid tumors growing from connective tissue, cartilage, bone, and muscle. They account for most cancers studied in laboratory animals, but for only two per cent of human cancers. Lymphomas, constituting five per cent of human cancers, involve excessive production of lymphocytes by the spleen and lymph nodes. Hodgkin's disease is one kind of lymphoma. Leukemias, constituting four per cent of human cancers, exhibit excessive production of leukocytes, the white blood cells.

All cancers are characterized by hyperplasia, the excessive production of cells "without cause." Although hyperplasia occurs in the healing of a skin wound, it stops at the wound's boundaries. Even in tissue culture normal cells grow only enough to form a thin turf in the dish or flask, being inhibited from further growth by cell-to-cell contact. Cancer cells in culture, however, are not contact-inhibited; they grow up and over each other. Cancer cells also function abnormally; for example, the excess lymphocytes in lymphoma are inactive immunologically. Finally, cancers have the ability to metastasize, that is, to spread: the cells at one site detach, migrate, and start a new cancer elsewhere in the body. (Overpopulation triggers migration.) Cancers that have not begun to metastasize can often be cured by irradiation, drugs, or surgery; but after metastasis treatment is usually unsuccessful.

While cancers produce no toxic substances, they kill the patient by nourishing themselves at the expense of his body, lowering the body's resistance to infection, crowding out essential tissues, or by other means. And though cancer is primarily a disease of old age, its frequency is increasing faster than population growth and life expectancy are increasing, according to Suss and his colleagues (1973).

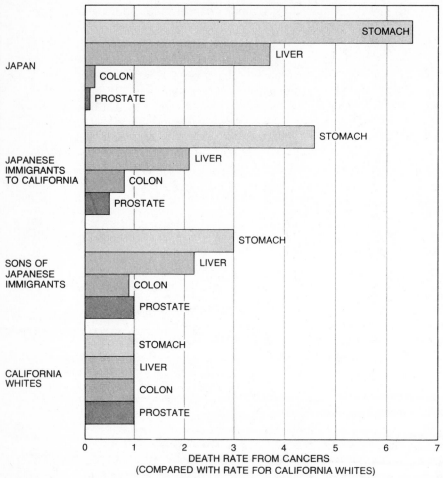

Figure 12–4 Differences between Japanese and whites in cancer incidence under various circumstances. (From J. Cairns: The cancer problem. Scientific American, 232: 64–77, 1975. Copyright © 1975 by Scientific American.)

This would seem to indicate that the causes of cancer are increasing in society.

What are these causes? Only some of them are known. Many researchers feel that environmental agents are major factors. Cigarette smoking is associated with lung cancer, but cats and dogs are also prone to lung cancer; this suggests an association with air pollution, which may indeed be important since lung cancer is more common in cities than in the countryside. Workers in certain industries — for example, asbestos workers and uranium miners — are especially subject to carcinogens, that is, cancer-inducing substances. But environmental pollution by carcinogens (e.g., x-rays and pesticides like DDT and dieldrin) affects us all.

Another cause of cancer may be one's type of diet. The carcinogen benzpyrene (actually 3,4-benzpyrene) is even found in charbroiled steaks. Dietary factors might explain why Japanese are six times more prone to stomach cancer than U.S. whites (Fig. 12–4). In addition, virus infections, immunological defects, emotional stress, and genetic predisposition play a role in cancer induction (as discussed subsequently).

Cancer has been studied for over a hundred years. By 1915 two Japanese researchers had shown that tar, which was already known to cause cancer in English chimney sweeps, would cause cancer in rabbits when it was painted on their ears.

Benzpyrene is one of the substances in tar. We now know that it is also a part of automobile exhaust, cigarette smoke, some smoked foods, and petroleum residues. But other substances similar to benzpyrene are not carcinogenic, or are only weakly so. The reason for this difference seems to be that benzpyrene is reactive in more ways with body constituents than are related types of molecules. Benzpyrene reacts with proteins in vivo, and it also reacts with DNA. Moreover, the human body has enzymes which activate some chemicals to a carcinogenic state, often by simply adding or removing a group such as CH_3. Recent work by Bruce Ames and others has amply demonstrated this activation.

Hypotheses About Cancer

Although there are many hypotheses to explain what triggers a cell to change to a cancerous state, we will consider here only those hypotheses centering on the immune system, cellular genes and mutation, and viral genes.

The defective immunity hypothesis proposes that through random mutation of lymph cells, or through the triggering action of carcinogens (which are known to be immunosuppressive), the body's process of immune surveillance of foreign matter (including new tumor cells) becomes ineffective. Although some tumors develop tumor-specific antigens which evoke a typical immune response in the body, the number of immune lymphocytes may be too low to block the development of a tumor. Whether this is due to mutation of lymph cells, emotional stress, poor nutrition, or a combination of these and other factors is unclear.

Another hypothesis proposes that a cancer cell's own DNA is altered. Let us consider the known facts. Item 1: Some strains of laboratory animals support tumor growth and others do not. Strains of mice have been bred which develop tumors in all animals. Humans with a genetic condition called xeroderma pigmentosum also develop skin cancers "spontaneously." This disease is a recessive autosomal one. What is inherited, though, is an oversensitivity of the skin to ultraviolet (UV) light (caused by a defect in an enzyme that repairs damage in DNA). If an individual with xeroderma stays in the dark, he or she does not develop cancer. It is not the cancer that is inherited here, only the predisposition to have UV-damaged DNA left unrepaired. This may be the case for the cancerous mouse strains too. In fact, mating experiments on high × low cancer strains indicate that one or two genes govern susceptibility of cells to tumor formation (Heston, 1974). Item 2: Most (but not all) cancerous tissues have abnormal chromosome complements, often with different numbers of chromosomes in different cells inside one tumor, although a recognizable "stem line" made up of both normal and abnormal (marker) chromosomes can usually be found in all the affected cells (Jones, 1974). Item 3: In patients with chronic myeloid leukemia, the white blood cells have a shortened chromosome No. 22 (the Philadelphia chromosome) and an elongated chromosome No. 9. (Presumably a piece broken off from 22 has translocated to 9.) Other cell types of the patients do not have these abnormalities. Patients with an eye cancer (retinoblastoma) are missing a middle segment of chromosome 13.

So far the evidence for DNA mutations or chromosomal alterations causing cancer does not look strong. Item 1 may indicate inherited vulnerability to certain environmental agents. Items 2 and 3 could be accompaniments or consequences of cancer rather than prerequisites for it. But let us go on. Item 4: Somatic mutations (chromosomal or genic) occur more often as individuals age, whether they contract cancer or not. Some of these mutations may induce cancers. In support of this notion is the finding that many carcinogens—x-rays, UV light, nitrosamines, benzpyrene, etc. — are also mutagenic. In any event, a tumor cell does pass on its cancerous character to its daughter cells. Perhaps several mutational defects are needed to start a cell on the road to cancer disaster (multi-hit initiation). Alternatively, carcinogens may act not by damaging DNA but by inducing "forbidden" gene transcription (for enzymes which stimulate mitotic events); that is, carcinogens may not be somatic mutagens but somatic inducers of DNA information which is normally repressed. Carcinogens may also be repressors of certain normal gene activities in precancerous cells. This idea is compatible with the finding that many carcinogens are known to bind to or combine with DNA. Granted that this binding of carcinogens to DNA may also be used as evidence in favor of the mutagenic hypothesis, the possibility that tumor cells become disorganized or unregulated in their decoding of DNA information has not yet been ruled out.

Are viruses the cause of human cancer? The wart virus is the only one definitely known to cause a human cell growth problem. But warts are benign, self-limiting growths. Although both DNA and RNA-DNA viruses have been shown to cause cancers in experimental animals and to cause cell transformation in tissue culture, the only solid indication so far that viruses cause cancers in people is the isolation of RNA-DNA virus-like particles from cells of patients with breast cancer and, more recently, from cells of patients with acute leukemia (Maugh, 1975). In both cases, however, the virus particles were not observed in tissues fresh from the patient; they were only seen as products of these tissues after they had been cultured in the laboratory. These isolated viruses could be the products of proviruses liberated from their chromosomal attachment (see Fig. 12–2). But H. E. Temin points out that particles similar to RNA-DNA viruses can also be isolated from noncancerous tissues, even from fetal tissues, and such particles are not found in cells from the majority of cancers. Temin therefore thinks it unlikely that RNA-DNA viruses are a cause of human cancer. A connection between viral and carcinogenic agents may prove to be that they both break chromosomes. Chromosomal breakage may or may not lead to cancer, depending on the precise location of the break and the fate of the unhealed chromosome ends.

SUMMARY

Viruses are diverse in their genetic makeup and in their ability to invade various organisms. Being parasites, they usually cause bodily harm to the host organism. Being very small, they are sometimes undetected contaminants of cell preparations in the laboratory. The simplicity of their anatomy and their nomadic behavior make them prime suspects as the elusive cause of human cancer. Other causes of human cancer are, however, known or conjectured; these include both environmental pollutants and malfunctioning genes.

STUDY QUESTIONS

1. How do cellular DNA replication and RNA replication of the polio virus differ?
2. Several types of tumors are inherited as dominant conditions (retinoblastoma, neurofibromatosis, multiple polyps of the colon). How might these phenotypes come about?
3. Cancer in humans has been reported in rare cases to undergo remission (be cured) spontaneously. By what mechanism might this happen?
4. Key terms:

   virus   provirus
   phage   hyperplasia

lytic cycle (productive infection)
lysogeny
vaccination
nonproductive infection
(cell) transformation
RNA-DNA virus

reverse transcriptase
contact inhibition
cancer
metastasis
carcinogen
xeroderma pigmentosum

REFERENCES*

Heston, W. E. Genetics of cancer. Journal of Heredity 65:262–272, 1974.
Jones, K. W. Chromosomes and malignancy. Nature 252:525, 1974.
Maugh, T. H. Leukemia: a second human tumor virus. Science 187:335–336, 1975.
Suss, R., V. Kinzel, and J. D. Scribner. Cancer: Experiments and Concepts. New York: Springer-Verlag New York, Inc., 1973.

SUGGESTED READINGS

Amber light for genetic manipulation. Nature 253:295, 1975.
Ames, B., W. Durston, E. Yamasaki, and F. Lee. Carcinogens as mutagens. Proceedings of the National Academy of Sciences 70:2281–2285, 1973.
Berg, P., D. Baltimore, S. Brenner, R. O. Roblin, and M. F. Singer. Asilomar conference on recombinant DNA molecules. Science 188:991–994, 1975.
Cairns, J. The cancer problem. Scientific American 232:64–77, 1975.
Chargaff, E. On the dangers of genetic meddling. Science 192:938–940, 1976.
Cohen, S. N. The manipulation of genes. Scientific American 233:25–33, 1975.
Comings, D. E. A general theory of carcinogenesis. Proceedings of the National Academy of Sciences 70:3324–3328, 1973.
Eckhart, W. The 1975 Nobel Prize for physiology or medicine. Science 190:650, 712–714, 1975.
Holland, J. J. Slow, inapparent and recurrent viruses. Scientific American 228:33–40, 1974.
Judson, H. R. Fearful of science. Harper's Magazine, June, 1975, pp. 70–76. Reports on a conference on DNA hybridization.
Kier, L. D., E. Yamasaki, and B. N. Ames. Detection of mutagenic activity in cigarette smoke condensates. Proceedings of the National Academy of Sciences 71:4159–4163, 1974.
Kolata, G. B. Phage in live virus vaccines: are they harmful to people? Science 187:522–523, 1974. See also Letter, Science 188:8, 1974.
Marx, J. L. Tumor immunology I: the host's response to cancer. Science 184:552–556, 1974.
Marx, J. L. Molecular cloning: powerful tool for studying genes. Science 191:1160–1162, 1976.
Maugh, T. H. Influenza: the last of the great plagues. Science 180:1042–1044, 1973.
Maugh, T. H. What is cancer? Science 183:1068–1069, 1974.
Sinsheimer, R. Troubled dawn of genetic engineering. New Scientist 68:148–151, 1975.
Spector, D. H., and D. Baltimore. The molecular biology of poliovirus. Scientific American 232:25–31, 1975.
Temin, H. M. RNA-directed DNA synthesis. Scientific American 226:24–33, 1972.
Temin, H. M. The cellular and molecular biology of RNA tumor viruses. Advances in Cancer Research 19:48–105, 1974.
Wade, N. Genetics: conference sets strict controls to replace moratorium. Science 187:931–935, 1975.
Wade, N. Recombinant DNA. Science 190:1175–1179, 1975, and 191:834–836, 1976.
Watson, J. D. Molecular biological approach to the cancer problem. In The Biological Revolution. W. Fuller (ed.). Garden City, New York: Doubleday & Co., Inc., 1972, pp. 169–182.
Watson, J. D. When worlds collide: research and know-nothingism. BioScience 23:438, 1973.

*Note: items cited in the References are also recommended for suggested reading.

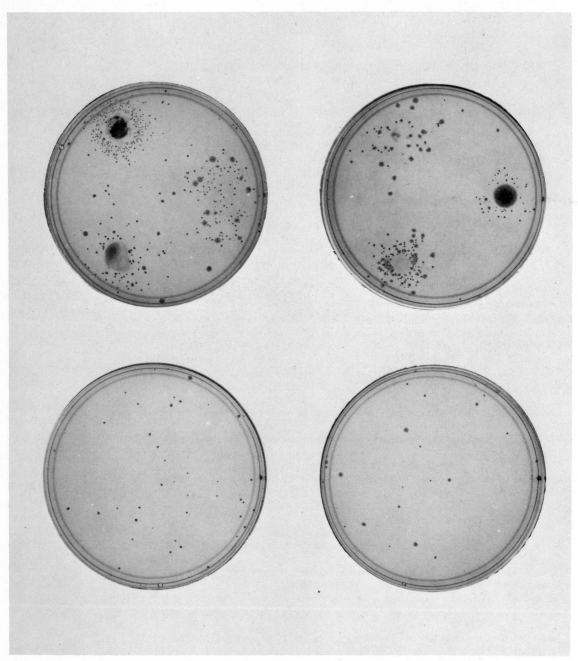

Spot tests on petri plates showing the mutagenicity of various hair dyes on bacteria: Moonhaze no. 32 (Miss Clairol), Moonlit Mink no. 360 (Clairol Born Blonde), Wild Fire no. 32 Roux fancitone) (all tested without peroxide), and Frivolous Fawn no. 23 (Roux fancitone) (mixed with H_2O_2 as per instructions). Control plates without hair dyes are on the right. Plates B, but not A, contain liver microsomes (S–9 Mix). Smokey Ash Brown no. 775 and Natural Black no. 83 (Clairol Loving Care) are semipermanent, non-oxidative type dyes. (From Ames, B. N., Kammen, H. O., and Yamasaki, E.: Hair dyes are mutagenic. Proc. Nat. Acad. Sci., *72*:2423, 1975.)

146

Chapter Thirteen

MUTATION

Heritable changes in the genetic material, whether spontaneous or artifically induced, large or small, are called mutations. Mutations are abrupt and generally rare in nature, but they are known to occur in many genes randomly over time. In previous chapters we saw how mutations in body cells might be related to the debilitating effects of autoimmune disease, aging, and cancer. As far as is known, the same mutational processes occur in mitotic (somatic) cells and in meiotic (germ) cells. But somatic mutations are passed on only to descendent cells in the body of the organism during that one generation, whereas germinal mutations (those in pre-reproductive or reproductive cells) can be inherited by descendants through many succeeding generations.

EVOLUTIONARY SIGNIFICANCE OF MUTATIONS

The basis of all genetically adaptive change in life is mutation. Mutation is the "raw material" on which genetic recombinational processes work. Although most mutations have maladaptive effects, some prove to be advantageous for their bearers in terms of survival and reproduction in certain settings. These advantageous mutations are selected for and perpetuated in the species until such time as the environment changes in ways which favor other alleles, perhaps ones due to subsequent mutations.

For many genes a species will be virtually monomorphic, almost all individuals being homozygous for the same alleles. This is interpreted to mean that a particular allele and its genic function is strongly selected, and that variation in that function is only barely, or not at all, tolerated. The majority of genes probably code for regulatory proteins, about which we know very little. Among the genes we do know well, many seem to code for enzymatic proteins that have been selected over countless generations for their ability to function well. Throwing a "monkey wrench" at random into this genetic machinery is unlikely to improve its efficiency. Thus the variant alleles of monomorphic enzyme-coding genes found in populations are usually lethal alleles or others that are selectively disadvantageous. As new mutations occur sporadically, they are culled. The frequency of disadvantageous variants, then, depends basically on the rate of occurrence of new mutations.

For other well known genes in a species, however, variant alleles are found rather frequently. These are called polymorphic genes. We have already discussed some instances of human polymorphisms: ABO blood groups, sickle-cell hemoglobin, and others. The variant alleles of these genes are too frequent to be accounted

147

for by new mutations alone, although the alleles originated by mutations, of course. Often the alleles are maintained in the population because of their different adaptive advantages in particular environmental situations, as is the case for the hemoglobin S and A alleles (see Table 10–1). In other situations the different alleles may simply be tolerated in the population, if all are compatible with survival and reproduction under the prevailing circumstances (selective neutrality). This may be the explanation for the blood group polymorphisms discussed in Chapter Eleven. We will return to a consideration of mutations in biological evolution in later chapters. The important point to keep in mind now is that mutations provide the genetic opportunities for adaptive change to unpredictable environments down through the generations of life.

CAUSES OF MUTATION

While mutations occur spontaneously, that is, without any known cause, they can also be induced experimentally by some chemical and physical agents. The agents that cause genic, or point, mutations generally also cause chromosome breaks, but not necessarily with the same degree of effectiveness. At the level of the gene, a consideration of DNA structure provides a basis for classification of point mutations into two types: base-pair substitutions and frameshifts.

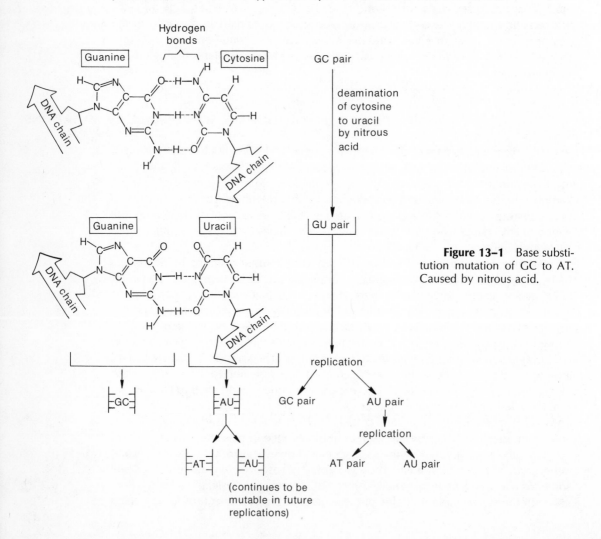

Figure 13–1 Base substitution mutation of GC to AT. Caused by nitrous acid.

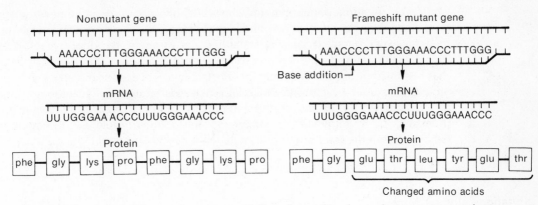

Figure 13–2 Frameshift mutations cause a shift in the reading frame during protein synthesis.

A base-pair substitution is a change in a DNA base pair, for example, from AT to GC. Base substitutions can readily be induced in intact cells, as well as in DNA replicating in vitro, by certain chemicals: base analogs like 5-bromouracil and deaminating compounds like nitrous acid. Base analogs are purines and pyrimidines that mimic the shape and hydrogen-bonding properties of A, T, G, or C. Base analogs incorporated into DNA can mispair during DNA replication, though, and thus produce DNA base substitutions. Many base analogs are known, including some in our daily diet such as the caffeine in coffee and tea. Caffeine is an analog of adenine and has been shown to cause mutations in cultured human cells. But it may not be a significant mutagen for human reproductive cells because, like many other ingested chemicals in the gut, it may not reach the gonadal tissue. Cells along the digestive track, however, might not be as safe from its effects. Definitive research on caffeine is needed.

Chemicals which can add or remove reactive groups from DNA bases have also been shown to cause base substitution mutations in laboratory studies. For example, nitrous acid can remove (deaminate) the amino group (NH_2) from cytosine, a reaction which sometimes results in the change of a GC pair to an AT pair in the subsequent generations of DNA (Fig. 13–1).

Frameshift mutations, which shift the reading of mRNA codons for translation into protein, can be induced experimentally by compounds such as acridine orange, a fluorescent dye. The flat dye molecule is thought to sequester itself in the DNA ladder as an extra "rung." Whether or not this occurs, the dye can cause the addition or deletion of a base pair in the DNA during subsequent replication. Since mRNA is decoded on ribosomes by tRNA molecules reading the message three bases at a time from a fixed starting point, the addition or deletion of a base produces a shift in the reading frame on the mRNA, thus scrambling the protein translation from the mutated site onward to the end of the message (Fig. 13–2). Such a frameshift mutation makes a "gibberish" protein, one incapable of functioning. Areas of the DNA which for one reason or another have monotonous runs of base pairs $\left(\dfrac{AAAA}{TTTT}\right)$ seem to be especially prone to frameshift mutagenesis. These areas are called "hot spots."

The causes of spontaneous mutations are largely unknown, although mistakes in DNA repair and replication have been implicated. The proportion of new mutations observed in humans that are spontaneous, rather than induced by identified environmental agents, is also unknown, but disputed. We can, however, point to substances in human environments that *probably* account for some new mutations in human beings; at least we know that these substances induce mutations in cells in experimental laboratory situations. For instance, the drug hycanthone induces both frameshifts and chromosomal abnormalities in cultured cells. This drug has been widely used in treatment for schistosomiasis, a disease of Asian and tropical countries caused by parasitic flukes. Compounds related to hycanthone are thus being screened for an antischistosome effect coupled to a lack of mutagenicity for human cells. By minimizing human exposure to known mutagens it may be possible to lessen the load of harmful mutational effects on populations.

Mutation Pollution

Most chemicals in the human environment are not mutagenic. The environmental mutagens best studied for human susceptibility are carcinogens, chemical pollutants, food additives, drugs, cosmetics, and radiation. It is not known for any of these chemical categories if the mutagens in question reach human gonadal tissues and cause mutations in the reproductive cells. But the chemicals can certainly reach somatic cells and produce mutations there. Radiation can and does reach reproductive cells; this will be discussed after we consider chemical mutagens more fully.

For obvious reasons people cannot be experimented on with possibly harmful substances. Mutagenic effects on humans are thus ascertained indirectly, by extrapolation from studies on human cells in culture, on mice, or other animals.

In the previous chapter some carcinogens were identified as mutagens. Many of the mutagenic carcinogens cause frameshifts, but a few cause base substitutions. Carcinogens usually become mutagenic, however, only after modification of their chemical structure by enzymes present in living cells. Preliminary observations suggest that individuals differ in their capacity to induce certain enzymes that can modify carcinogens. Thus some people may be genetically more susceptible than others to the cancer-causing and mutation-causing properties of these compounds.

Certain people are exposed to mutagens more often than is the general public. Not only do individuals in particular occupations risk overexposure, but segments of the population taking certain drugs (e.g., hycanthone, isoniazid) are also special targets for mutations. Isoniazid, regularly used in the treatment of tuberculosis, has been shown to be carcinogenic in mice and mutagenic in cultured cells. Geneticists have urged that many drugs in wide use, such as aspirin, be thoroughly tested for possible mutagenic activity. But "megamouse" experiments, the preferred kind of test system for screening possible mutagens, are enormously expensive (about $100,000 per test). Moreover, mice are not little people: a substance mutagenic in humans might be labeled nonmutagenic on the basis of tests employing the metabolic system of mice. In addition, there is the problem that over a million chemicals are known, and of the one to two thousand that are possibly human mutagens, no more than several hundred have been tested rigorously for mutagenicity. The task ahead is enormous, and, because of the continued development of new synthetic chemicals for use in industrial processes, cosmetics, textiles, and so on, the task is probably endless.

Among agricultural chemicals, the pesticides dieldrin, aldrin, and DDT have all been shown to be mutagenic. Their use in the U.S. has now been banned or restricted by governmental action (although the soil in Illinois corn fields, and presumably those of other states, already consists of 1 per cent dieldrin, a high level of contamination).

Food additives that are mutagenic include the sweetener cyclamate of diet-drink fame (now under government restriction) and nitrites. Nitrites (NO_2) are used as food preservatives and enhancers of flavor and color in meat (such as hot dogs). U.S. adults consume only about one fifth of an ounce (4.5 grams) of nitrites a year; although it might seem like a small quantity, this is a significant amount of internal pollution. Exactly how nitrites react inside the body is not known, but they are suspected of forming nitrous acid (HNO_2) in some circumstances (see Fig. 13–1), and also of combining with amines (NH_3) to form nitrosamines, known to be carcinogens and mutagens in their own right.

In 1974 Bruce Ames discovered that the majority of commercial hair dyes sold in the U.S. are mutagenic to cells in culture. Some of the dyes contain 18 different chemicals, 9 of which act as mutagens. Some of the dye chemicals are absorbed through the scalp and can subsequently be detected in the urine. Although 20 million people in the U.S. dye their hair, hair dyes are currently exempt from governmental controls by action of Congress taken in 1938. The cosmetic industry felt that hair dyes should be exempt because of the huge demand for them and because they could not meet safety standards of the Food and Drug Administration. Congress responded accordingly.

Vinyl chloride is a gas used to make the plastic in phonograph records, upholstery textiles, and floor tiles (polyvinyl chloride). Until recently it was also used as a propellant in aerosol sprays. Some workers in factories using vinyl chloride have developed liver cancer, presumably from exposure to it. Vinyl chloride demonstrably causes liver cancer in laboratory animals, and, according to Bruce Ames, causes mutations in cells in culture. It is a common chemical additive in consumer products. Five billion pounds of polyvinyl chloride were produced in 1973 in the U.S.; the Environmental Protection Agency estimates that 200 million pounds of vinyl chloride escape each year into the American environment as a result of its industrial usage. The plastics industry says a ban on vinyl chloride would put 2.2 million people out of work, according to R. Gillette in 1974.

The evidence for LSD and marihuana as mutagens, mentioned in Chapter One, pales in comparison with that for the industrial chemicals noted here. Not only is the evidence inconclusive that LSD and marihuana induce mutations, but the entire population is not exposed to them, as it is to vinyl chloride, food additives, and pesticides.

Radiation

Radiation is electromagnetic waves of any wavelength and includes gamma rays, x-rays, ultraviolet light, visible light, and even radio waves. X-rays and gamma rays are mutagenic forms of radiation. Their short wavelengths have considerable ability to penetrate living tissue. Visible light has relatively long wavelengths and is not mutagenic. Ultraviolet light is intermediate in wavelength and in humans only penetrates as far as the layers of the skin, where it causes sunburn and skin cancer, presumably by somatic mutation. Ultraviolet light is, however, an effective mutagen in the sparsely layered cell cultures studied in the laboratory. Like visible and UV light, gamma rays fall on the earth from cosmic sources. X-rays are produced by machinery. Other sources of radiation are radioactive rocks and soils (such as the uranium ores mined on the Colorado Plateau and used as fuel for nuclear reactors).

"I figure, what's it all matter? If cigarettes don't get you, radiation will."

Some of this radioactivity we ingest with food and drinking water, thus internalizing the sources of radiation.

The mutagenic effect of x- and gamma radiation results from their ability to penetrate into cells and to upset the electron balance of cellular molecules. They are thus said to be ionizing radiation. Their high-energy waves may also directly break DNA and chromosomes into pieces, both in somatic and germ-line cells. Ultraviolet radiation is nonionizing and is of little importance in human germ-line mutagenesis because it penetrates poorly into tissues.

The unit of radiation dose absorbed by human tissue is called the rem. It takes into account the efficiency—or penetrating power—of the radiation being discussed. X-rays currently used in medical practice give, on the average, a dose of radiation of 0.5 rem per x-ray film. Of this 0.5 rem only about 14 per cent (0.07 rem) reaches reproductive tissues, the gonads. The gonadal dose from chest x-rays is the result of scatter of x-rays from the machine.

Whole-body radiation of 300 rem will kill about half of the human population so exposed. It does not kill uniformly because of biological variation (health, genes,

TABLE 13–1   Estimates of Whole-Body Radiation in the U.S. Population*

Source	Average Dose Rate (rem/yr.)	Proportion of Exposure
Natural radiation	.102	56% of total
Man-made radiation:		44% of total
Fallout from atomic tests	.004	5% of man-made
Nuclear power industry	.000003	.004% of man-made
Medical-dental (based on abdominal dose)	.073	91% of man-made
Other (occupational and miscellaneous)	.003	4% of man-made
Total	.182	

*From Table 2, BEIR (Advisory Committee on the Biological Effects of Ionizing Radiations), 1972, p. 19.

and other factors). Many people received such high doses of radiation in the 1945 atomic bomb blasts at Hiroshima and Nagasaki.

The folly of man does not usually go to such lengths, however. In the U.S. we have some radiation exposure to contend with, about 0.18 rem per person per year on the average (Table 13–1). Of this, 0.10 rem is from natural sources (cosmic rays, internal and terrestrial radiation), amounting to about 3 rem per 30 year period (the average generation time). The other major source of radiation exposure in the U.S. is medical diagnostic x-rays, of which we average 0.07 rem per year, or about 2 rem per generational period. Of this, 79 per cent, or 0.055 rem per year, is the gonadal (genetically significant) dose. People in Europe, and elsewhere, receive far less medical x-ray exposure, a condition which the medical profession in the U.S. should seek to achieve.

A U.S. government report (called the BEIR Report) has suggested that it is possible to achieve a 50 per cent reduction in radiation exposure from medical sources by: (1) stopping repetitive x-ray tests; (2) routinely using lead aprons to shield areas of the body, especially the pelvis, for which irradiation is unnecessary; (3) improving test machinery and techniques; and (4) training x-ray personnel in new techniques. The 50 per cent reduction refers to the genetically significant dose, that is, that dose of irradiation which reaches the gonads and can cause germ-line mutations. This exposure is most important in terms of long-range mutational damage to succeeding generations. The genetically significant dose for natural radiation is estimated to be 88 per cent of the 3 rem per generation we receive (2.6 rem). For medical diagnostic radiation the genetically significant dose is estimated to be 41 to 82 per cent of the 2 rem per generation we receive.

But, besides causing germinal mutations, radiation is known to cause somatic mutations and to be carcinogenic. Among children aged two years and older who contract acute leukemia, some are known to have had substantial radiation exposure in utero. Deaths from leukemia and other forms of cancer are about 40 per cent higher for those exposed in utero than for those not exposed. The latent period for the development of childhood leukemia is as little as three years. (For other cancers the latent period is much longer, perhaps 10 to 40 years: children who received high dose x-ray treatment for swollen tonsils or adenoids from the 1930's through the 1950's are now showing a high incidence of thyroid cancer. Programs to locate adults who received the x-ray treatment in childhood are underway in several cities in the U.S.)

Radiation effects, both mutagenic and carcinogenic, have been shown to be cumulative in a variety of studies on laboratory animals. The more radiation, the more mutations and cancers produced. This is an important point to bear in mind. Another point of major significance for human health is that acute radiation exposure is much more harmful than the same dose of radiation received over a long period (chronic exposure) because the body can partially repair radiation effects if given sufficient time. For example, a dose of 3 rem, received all at once (acutely), is more likely to produce mutagenic and other bodily damage than a 3 rem dose received over a normal life span.

MUTATION RATE

Achondroplasia is a dominantly inherited form of dwarfism. Since these dwarves usually die before adulthood, new cases are almost always the result of new germinal mutations. In a study of 94,000 births to nondwarf parents 8 achondroplastic children were found. Since each child is the product of two gametes, this sampling represents a mutation rate of 8 mutant alleles per 188,000 gametes, or 1 mutation per 23,000 gametes. On the basis of this and similar studies the human mutation rate for an average gene is estimated to be 1 mutation per 100,000 gametes per generation. The mutation rate for aneuploidy of all kinds is much higher, you may recall (Chap. Six); it is perhaps 1 mutation per 100 gametes per generation. Because of the repetitiousness of mutations it has been estimated that we all carry 3 to 5 new recessive, harmful mutations. But because of the randomness of mutation, we are (probably) all mutant in different ways.

On the basis of irradiation experiments on laboratory animals there is no doubt that radiation can cause germinal mutations in the human species, even though studies on the Japanese atomic bomb survivors have not yet disclosed any germinal effects. (But the techniques used to study the Japanese survivors were relatively crude and few people of child-bearing age have been studied thus far; remember that recessive mutations may not be revealed until later generations.)

It has been estimated that the "doubling dose" of radiation for inducing mutations is between 20 and 200 rem. The 200-rem figure is preferred by a number of geneticists. This means that a radiation dose of 200 rem will double the spontaneous mutation rate in humans from, say, 1 mutation per 100,000 gametes to 2 per 100,000 gametes. Barring atomic warfare, the human population is obviously not in danger of such an increase in the mutation rate due to radiation exposure. But, given the harmful nature of most mutations, even a small increase in the mutation rate, whether due to radiation or chemical factors, can cause untold human suffering.

SUMMARY

Mutations occur both spontaneously and by the action of physical and chemical agents. Most mutations are harmful although some provide the raw material for adaptive evolutionary change. The pollution of human life by mutagenic chemicals· and radiation in technological countries is alarming.

STUDY QUESTIONS

1. If you were the administrator in charge of a government laboratory for cancer research, what kind of research might you emphasize or concentrate on?

2. Do you agree with the last sentence in the Summary? If not, why? If so, what do you think can be done about the situation?

3. Key terms:

germinal mutation	base-pair substitution	radiation
somatic mutation	base analog	rem
mutagen	frameshift	mutation rate

SUGGESTED READINGS

Advisory Committee on the Biological Effects of Ionizing Radiations (BEIR). The Effects on Populations of Exposure to Low Levels of Ionizing Radiations. Washington, D.C.: National Academy of Sciences–National Research Council, 1972.

Ames, B. N., H. O. Kammen, and E. Yamasaki. Hair dyes are mutagenic. Proceedings of the National Academy of Sciences 72:2423–2427, 1975.

Challis, B. C., and C. D. Bartlett. Possible cocarcinogenic effects of coffee constituents. Nature 254:532–533, 1975.

deSerres, F. J., and W. Sheridan (eds.). The Evaluation of Chemical Mutagenicity Data in Relation to Population Risk. Environmental Health Perspectives, No. 6. Department of Health, Education, and Welfare, No. (NIH) 74–218, December, 1973.

Drake, J. W. The Molecular Basis of Mutation. San Francisco: Holden-Day, Inc., 1970.

Ducatman, A., K. Hirschhorn, and I. J. Selikoff. Vinyl chloride exposure and human chromosome aberrations. Mutation Research 31:163–168, 1975.

Environmental mutagen hazards. Science 187:503–514, 1975.

Gillette, R. Cancer and the environment. II: groping for new remedies. Science 186:242–245, 1974.

MacMahon, B. Prenatal X-ray exposure and childhood cancer. Journal of the National Cancer Institute 28:1173, 1962.

Maugh, T. H. Chemical carcinogenesis: a long-neglected field blossoms. Science 183:940–944, 1974.
    Discusses enzyme induction for modifying carcinogens to produce lung cancer.

Miller, R. W. Delayed radiation effects in atomic-bomb survivors. Science 166:569–574, 1969.

Nader, C. The dispute over safe uses of X rays in medical practice. Health Physics 29:181–206, 1975.

Roth, J. R. Frameshift mutations. Annual Review of Genetics 8:319–346, 1974.

Seligmann, J., and M. Butler. Dangerous legacy. Newsweek, April 14, 1975, p. 82.
    Discuss the thyroid center location program.

Shapley, D. Nitrosamines: scientists on the trail of prime suspect in urban cancer. Science 191:268–270, 1976.

Wolff, I. A., and A. E. Wasserman. Nitrates, nitrites, and nitrosamines. Science 177:15–27, 1972.

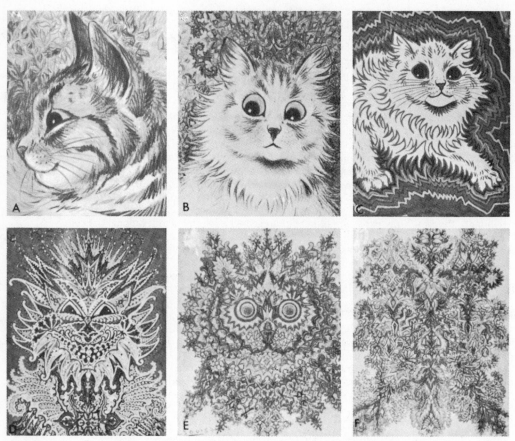

A series of paintings made by an English Artist, Louis Wain, showing the progress of his schizophrenic breakdown. (© Guttmann Maclay Collection, Institute of Psychiatry, University of London.)

Chapter Fourteen

GENES AND BEHAVIOR

Behavior is a term with many meanings: social, emotional, sensory, "normal" vs. pathological. Behavior is an effect of the nervous system and its accessory sense organs, but is influenced greatly by the rest of our anatomy and physiology, as well as by our social world and other aspects of our environment.

On the basis of pedigree analysis showing Mendelian segregation, a single gene can sometimes be shown to be the cause of a behavioral trait. In single-gene disorders the behavioral effect is often secondary to a physiological effect. For instance, phenylketonuria (PKU) is caused by an enzyme upset that secondarily produces paleness of the skin and hair and, without dietary treatment, mental retardation (see Box on PKU, Chap. Ten). The behavioral effect of retardation is, we say, one of the pleiotropic effects of the PKU genotype.

The mental degeneration that occurs in children with Tay-Sachs disease (see Box on TSD, Chap. Ten) is a pleiotropic effect of the genotype that is closer to the primary gene action than is the case in PKU. In Tay-Sachs the fatty content of nerve fibers is abnormal because of an enzyme deficiency in lipid metabolism. Lesch-Nyhan disease is another trait with pleiotropic behavioral effects (see Box on Lesch-Nyhan Disease, below). In the case of Huntington's disease, a rare type of mental degeneration that affects adults, the substratum of gene action is unknown. Here we can only assume that, as with the disorders mentioned above, the mental defect is secondary to a simple metabolic problem. (See Box on Woody Guthrie and Huntington's Disease, p. 158.)

But not all mental disability is genic in origin; a blow on the head will do. There are also disorders associated with no known environmental insults and with no history of such disability in the family—conditions for which a cause cannot even be guessed at. In addition, there are the so-called multigenic (or polygenic) conditions of behavior disability; they are thought to be greatly influenced by environmental circumstances. Finally, there are the behavioral conditions, such as Down's syndrome, which are caused by chromosomal alterations.

LESCH-NYHAN DISEASE

Children with Lesch-Nyhan disease are mentally retarded, muscularly spastic, and have the bizarre characteristic of compulsive self-mutilation (even to the extent of literally eating their own fingers). These children usually die before adolescence.

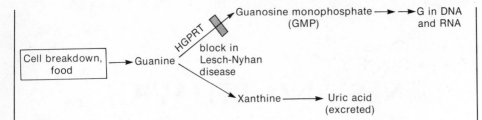

Lesch-Nyhan disease is a rare X-linked condition caused by the deficiency of an enzyme called HGPRT. HGPRT is necessary to salvage excess purines in the body for nucleic acid synthesis. The enzyme deficiency leads to an accumulation of uric acid in the urine—uric acid being an end product of the purine breakdown pathway prior to excretion (above). The uric acid build-up produces symptoms of gout (although gout is usually thought of as a disease of old people). Persons with gout, however, are not prone to self-mutilating behavior.

The reason why a deficiency of the HGPRT enzyme produces behavioral abnormalities is unknown, although several possibilities are being studied. Rats and mice with chemically induced behavioral symptoms similar to human Lesch-Nyhan provide one laboratory model for such studies.

Tests are now available to identify women who are carriers of the Lesch-Nyhan gene and also to diagnose the fetus in utero by amniocentesis to determine whether it has the Lesch-Nyhan disorder. The tests make possible effective genetic counseling (Chap. Fifteen) for families which are at risk for Lesch-Nyhan disease.

WOODY GUTHRIE AND HUNTINGTON'S DISEASE

Woody Guthrie was born in Oklahoma, grew up in the Dust Bowl of the Great Depression, and became a great balladeer and poet of the American spirit. His songs "So Long, It's Been Good to Know You," "This Land Is Your Land," and many others are his gifts to us. While young adults may know Woody best as the father of Arlo Guthrie, many members of the family have shown musical talent.

When Woody was 18 years old his mother died of an inherited, dominant disease called Huntington's chorea. It is characterized by a slow deterioration of the nervous system and choreic (spasmodic) movements. Afflicted individuals have a shortened lifespan. Woody Guthrie also died of this disease at age 55 in 1967. Each of Woody's children has a 50% risk of inheriting this disease as a result of Mendelian segregation: Aa (Huntington's phenotype) × aa (normal) → 50% Aa + 50% aa. The average age of onset of the disease is 35 years, but the range is from 20 to 50 years. It is to be hoped that Arlo and the other two living children of Woody and his wife Marjorie will not lose out in the genetic gamble on Huntington's. There is no effective treatment or cure for the disease.

In the early years of his illness Woody was intermittently depressed and irritable and became noticeably lopsided in his walk. Then in 1952 he had an attack of violent behavior and was hospitalized for several months, although his problem was not diagnosed. The Guthrie family found out Woody had Huntington's disease only after he suffered another outburst of violence and had become quite uncoordinated in all his movements. As a result of Marjorie Guthrie's pleading to know what was wrong with Woody, eventually a young doctor came up with the correct diagnosis.

This diagnosis caused some confusion. Although the family knew Woody's mother had died of "insanity," they found out it was Huntington's disease only after Woody was correctly diagnosed. Even then they thought it couldn't have been inherited by Woody because they were told the disease was only passed down to females. Such confusion reflects the lack of genetic expertise in the medical profession at that time (and the lack of genetic counseling referral centers, discussed in Chap. Fifteen).

The principal problem faced by the Guthrie family in Woody's last years was his need for institutionalization as he became unmanageably uncoordinated and also unable to speak. The family found no appropriate institution to care for victims of the disease. The only institutions available were for the mentally retarded or psychologically disturbed. This impasse still exists today for many victims of Huntington's disease although, thanks to the efforts of Mrs. Guthrie and others who have been involved with people suffering from Huntington's, the medical community is now better informed of the social services needed by these patients.

METHODOLOGY

The methods used by geneticists for studying behavioral traits include pedigree analysis, twin studies, adoption studies, and consanguinity ("same-blood") studies. Pedigree analysis has contributed to our understanding of only a few behavioral traits, those with clear-cut distinctions from the normal such as Huntington's and other rare diseases of the nervous system. For behavioral traits that differ from normal only by degrees, pedigree analysis has been less successful. The most common psychoses, schizophrenia and manic-depression, are among the traits which do not show a clean pattern of gene segregation. (See Box on Common Behavioral Problems, p. 160.) Sometimes failure in pedigree analysis is due to genetic heterogeneity in the disease where, say, an X-linked gene is at fault in some individuals and an autosomal one is at fault in others. Epilepsy has been shown to be a heterogeneous genetic condition by careful study of its transmission pattern in different families (Inouye, 1973), although not all cases of epilepsy have been shown to have a genetic basis. Pedigree analysis also is ineffective in explaining conditions caused by multiple gene combinations (polygenic conditions).

SOME COMMON BEHAVIORAL PROBLEMS

Epilepsy.   Disorder characterized by periodic convulsive attacks.

Mental retardation.   Generally defined to include persons with I.Q. scores below 70. Many causes. Over 2% of the U.S. population is retarded; 70% of this retardation is due to Mendelian variant alleles, 15% to polygenic conditions, 15% to Down's and other chromosomal abnormalities, 5% to brain injury and infections, 15% to other environmental factors, and the remaining 43% to unknown causes (Stern, 1973). Not surprisingly, then, most retardates are born to nonretarded parents. Some retardates need supervisory care. Severely mentally retarded persons are those with I.Q. scores below 50; all of them need supervisory care.

Psychosis.   Any of the possibly severe mental disorders, for example, manic-depression and schizophrenia.

Schizophrenia.   A moderate to severe psychiatric disorder characterized by delusions, hallucinations, and other well-marked thought defects, as well as suspiciousness, withdrawal, and passivity. Probably affects about 1% of the population worldwide, but the definition is not clear-cut and thus psychiatrists may disagree on a diagnosis in borderline cases.

Manic-depression.   A moderate to severe psychiatric disorder characterized by extremes of affect and mood. Among the psychoses, only schizophrenia is more common in the U.S.

Identical twins are the products of a single zygote which splits into separate halves. They are thus called monozygotic (MZ). Fraternal twins arise from two fertilized eggs and are called dizygotic (DZ). Dizygotic twins are therefore no more alike genetically than are siblings born at different times. But they differ from two single siblings in having had a shared uterine environment and, usually, a more similar childhood environment.

Monozygotic twins form a two-person clone, that is, they are genetic replicates. The identical genotypes of MZ twins provide background for comparing the effects of different environmental influences on the same genetic makeup. A common method of twin studies is to compare a group of MZ and DZ twins as to the proportion of pairs that are alike in having a particular trait (concordance) vs. the proportion of pairs in which one has the trait and the other doesn't (discordance). Since MZ pair members are identical in genotype, whereas DZ pair members differ in 50% of their genotype on the average, any trait that shows a greater concordance among MZ twins than among DZ twins must have a substantial genetic basis. (This assumes that environmental differences between pair members are the same for MZ's and DZ's, an assumption challenged by some geneticists.)

Monozygotic twin pairs have been found to be more concordant than DZ twin pairs for a number of behavioral traits, suggesting a strong genetic "loading" for these conditions: schizophrenia, manic-depression, criminal imprisonment, homo-sexuality (only males studied), and degree of alcohol consumption. The ratio between concordance frequencies in MZ and DZ twins can be used to calculate what is called a heritability value for the trait in question. (Heritability is an estimate of the influence of heredity in the variability of the trait in a population.) For traits that are polygenic, heritability values can be estimated using techniques of statistics, such as correlation coefficients and analysis of variance. However, heritability values are of doubtful validity in human genetics because it is difficult to determine if the estimates are biased by unknown factors, and also because heritability is a concept that is often misunderstood and misused (see section on I.Q. in this chapter).

Do parents who are related to each other produce offspring with a given behavioral trait more frequently than parents who are unrelated? To answer this question the degree of relationship between parents is compared for the affected group under study (consanguinity analysis). For many behavioral traits consanguinity between parents is associated with a higher frequency of the trait in the family, indicating the effect of common genes or common environmental factors.

Another method used in behavior genetic analysis is the adoption study: here adopted children are compared with their biological parents (with whom they share genes) as well as with their adoptive parents (with whom they share only environments).

All these various types of studies are often used in complementary fashion to unravel the basis of complex behavioral traits. But even if genetic factors occur in such a trait, it does not mean that all cases of individuals showing the trait are genetic in origin, nor does it tell us what proportion of the cases do involve genetic causes. The methodology considered here only permits us to glimpse the broad outline of the genetic component in behavior.

SOME MAJOR PROBLEMS

Probably the most common psychosis in the U.S. is schizophrenia. The cause of this mental illness is unclear. Although it runs in families and has other attributes of a genetic condition, Mendelian inheritance in a large lineage has not been demonstrated. Among the many reasons that have been advanced to explain the fuzziness of the genetic picture of schizophrenia, the most important may be the variable age of onset of the disease. All data on risks of schizophrenia in families must be age-adjusted to give realistic appraisals of the likelihood of its occurrence, and such adjustments have not been done in most studies to date, according to Elston.

What is known about the genetics of schizophrenia may be stated in terms of the methodologies discussed above for studying behavioral traits: (1) close relatives of schizophrenics are affected more often than more remote relatives (consanguinity studies); (2) MZ twin pairs are more often concordant for it than are DZ pairs (twin studies); and (3) adoptive/foster parents of schizophrenic individuals are less often schizophrenic themselves than are the biological parents (adoption/fostering studies).

In a particularly well-designed foster-home study, Heston interviewed and tested the adult children of schizophrenic mothers who had been separated from their mothers within one month after birth and compared these "index cases" to controls (similarly adopted) with no psychiatric family history who were matched to the index cases pair-wise for age, sex, and other variables. Heston found that 17 per cent of the index cases, but none of the controls, were schizophrenic. He also found that there was a higher incidence of mental deficiency, sociopathic personality, felony, problem drinking, and other psychological problems among the index cases than the controls. Perhaps even more interesting is his finding of "significant musical ability" in four of the 47 index cases, as compared with zero of the controls, and, furthermore, that the 21 of the 47 index cases who had no psychological problems "...possessed artistic talents and demonstrated imaginative adaptations to life which were uncommon in the control group" (Heston, 1966, p. 824).

Heston later showed that when schizoid personality (a mild version of schizophrenia) was tabulated together with schizophrenia in consanguinity and other studies, a single dominant gene could be hypothesized as the major factor in these disorders (Fig. 14–1). Whether an individual with the dominant gene develops schizophrenia or merely a schizoid personality, according to this model, depends on environmental influences and the modifying effects of other genes.

If there is a dominant gene with major effect on schizophrenia, it should be possible to discover its biochemical disorder in affected individuals. But despite sporadic reports of chemical anomalies in the brain tissue or blood of schizophrenics, no such anomalies have yet been convincingly substantiated (Wyatt et al., 1975; Shaskan and Becker, 1975).

Treatment for schizophrenics is not ideal. Drugs in use (e.g., phenothiazines) do not rehabilitate patients to full functioning. The mode of action of such drugs is disputed. Treatment may improve in efficacy if a biochemical lesion in the brain tissue of schizophrenics can be unequivocally identified. However, to prevent the occurrence of schizophrenia, rather than just treat it once it has occurred, it would be worthwhile to determine the factors which differ in the environments or life styles of those MZ twins which are discordant for the disease. (Roughly 14% of MZ's are discordant and 86% concordant for schizophrenia.) Finding such environmental differ-

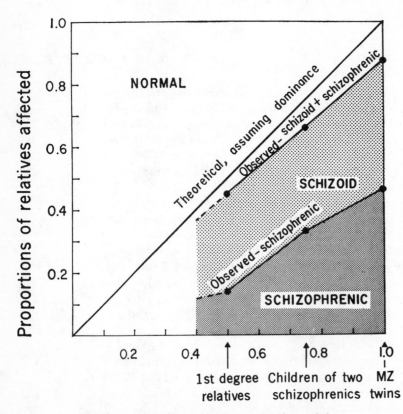

Figure 14–1 Observed and expected proportion of schizoids and schizophrenics among relatives. (From Heston, The genetics of schizophrenic and schizoid disease. Science, *167*, 249–256, 1970. Copyright © 1970 by the American Association for the Advancement of Science.)

ences between discordant MZ twins could have widely beneficial results, as it is less difficult to change our environment than our genes.

Extreme mood states of euphoria and depression, or sometimes depression alone, characterize the so-called affective illnesses. Psychic turmoil is at the center of these illnesses, the bipolar manic-depression and the unipolar depression psychoses. Depression can usually be traced to the disruption of an important human attachment—the loss of a mother by a child, or of a husband or wife by an adult. Depression may also be a reaction to withdrawal of personal approval by intimate friends or relatives. Certain drugs can artificially produce a depressive state; others can relieve depression. These chemicals are thought to cause, or alleviate, imbalances in the chemical triggers for nerve impulses in the middle part of the brain. The treatment of choice for manic-depression is the drug lithium. The reason for its effectiveness is not known. Lithium is a natural mineral in many rocks and is a component of mineral waters. Mineral springs have been used as general curatives in European countries since Roman times. (They are also used in the U.S., originally by the American Indians.)

All forms of affective psychoses are underlaid by genetic factors mediated in some way through the chemical derangements in nerve tissue. The specific gene-product abnormalities involved are unknown. Even in studies on induced depression in animals it is not known what the basic biochemical upsets in the brain are. It is also not known why the genetic vulnerability to affective illness shows up mainly in middle age. Most manic-depressives are in the age group of 30 to 50 years. Perhaps during these years there is more stress and a general feeling of greater hopelessness and helplessness (worsening health, decline in economic expectations, withdrawal of children, and so on). Possibly a mixture of chemical-genetic factors and psychological feelings of worthlessness combine to bring on affective illnesses.

Manic-depression is more common in Iceland (2–3% of the adult population) than anywhere else it has been studied. In New York State about 0.4% of adults are affected. Such differences could be due to genetic or climatic differences in the populations, or to diagnostic differences. Suggestive evidence for a genetic basis in manic-depression comes from the fact that the chance of manic-depression is high if one parent is manic-depressive (6–24% in different studies) and even higher if both parents are manic-depressives (20–40%). But here again the effects might be related to environmental factors, as the relevant studies are not all concerned with adopted persons. However, MZ twins are more often concordant for manic-depression (50–93% in different studies) than are DZ twins (0–38%). The risk of affective illness for first-degree relatives—parents, children, siblings—of manic-depressive patients is 39%. This points to the influence of a partially dominant gene. (If it were a fully dominant gene, the risk should be 50%). Further data suggest that the dominant gene is X-linked (Mendlewicz and Rainer, 1974).

Is there a relation between manic-depression and creative activity? Some manic-depressive patients discontinue their lithium treatment because they feel it stultifies their high-energy activity (which is maximal during the manic phase of the illness). A number of people in the creative arts are in this category of patients who reject lithium treatment.

Is there a significant behavioral effect of the XYY condition? The first reports on this chromosomal abnormality suggested that XYY's were prone to criminality, and this interpretation was given sensational coverage in newspapers and magazines. XYY's are unusually tall on the average and some are mentally retarded, but perhaps no more so than XY's are. Is it perhaps these attributes of XYY's which make them prone to imprisonment for criminal offenses? This may be the case, but XYY males have been found who are tall, retarded, and without any prison record.

The evidence in favor of XYY as a cause of criminality is of several sorts, of which two will be mentioned here. (1) Hook in 1973 reviewed various surveys for XYY's in prisons and found that approximately 2% of prisoners of all heights have the XYY condition. This 2% figure is greater than the frequency (0.1%) of XYY's among newborn males. (2) When one group of investigators studied the frequency of anti-social behavior among the brothers of XY and XYY prisoners, a higher frequency of such behavior was found among the brothers of XY's than among the brothers of XYY's. This result can be interpreted as meaning that the chromosomal imbalance for XYY's is more important in their antisocial behavior than is their family environment, whereas for XY males it is the family environment that is more important. But more careful studies on larger numbers of XYY prisoners—using controls matched for age, I.Q. scores, and other factors—later found no difference between the brothers of XYY and XY prisoners.

Taking into account these and other factors, a review by Boraonkar and Shah of all the available information on XYY's recently concluded that ". . . the frequency of antisocial behavior of the XYY male is probably not very different from non-XYY persons of similar background and social class" (p. 188).

Jonathan Beckwith and his colleagues (1975) have questioned recent hospital surveys of the chromosomal constitution of newborns. Since parents have been informed when their surveyed infant is found to have XYY (or any other disorder), Beckwith fears that the infants so identified may be stigmatized their whole life, or at least treated with some reserve by their parents, and thus become behavioral problems as a self-fulfilling prophecy.

How does this research square with our belief in a person's innocence until he or she is proven guilty? Is XYY a disease? Do XYY boys have rights that have been violated by their having been surveyed for chromosomal abnormalities as newborns? Beckwith states that such a survey ". . . cannot yield meaningful results, has no benefit but substantial risks to the families involved, and only serves to propagate the damaging mythology of the genetic origins of 'antisocial' behavior."

To put the XYY debate in reasonable perspective, one critic has commented that the XYY controversy ". . . is an example of our fascination for exotic problems to the neglect of common but more serious genetic conditions, such as the XY karyotype . . . Overwhelming statistical evidence indicates that the XY karyotype is associated with major social problems such as violent crime and war" (Mage, 1975).

QUANTITATIVE TRAITS

Human quantitative traits include height, weight, I.Q. score, number of children an individual produces, and many others. On the assumption that there is a genotype lurking behind every phenotype, it is generally conceded that there is some genetic basis for differences in these traits. But quantitative traits do not show clean Mendelian segregation ratios as do "yes-or-no" single-gene traits like PKU and albinism (Fig. 14–2). If, for example, all the children of 6-foot men married to 4-foot women grew to exactly 5 feet in height, and if all 5-foot × 5-foot matings produced 25% 6-foot, 50% 5-foot, and 25% 4-foot children, the height would fit a single-gene model with AA (6 ft.) × aa (4 ft.) → all Aa (5 ft.), and $Aa \times Aa \rightarrow \frac{1}{4}AA + \frac{1}{2}Aa + \frac{1}{4}aa$.

Obviously this does not occur. Five-foot couples have children that grow to a variety of heights, although short parents tend to have short children and tall parents tend to have tall children.

A. If adult height were due to a single gene with additive effects of the A allele (two examples):

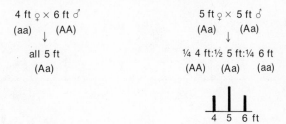

4 ft ♀ × 6 ft ♂
(aa) ↓ (AA)

all 5 ft
(Aa)

5 ft ♀ × 5 ft ♂
(Aa) ↓ (Aa)

¼ 4 ft:½ 5 ft:¼ 6 ft
(AA)   (Aa)   (aa)

4   5   6 ft

B. If height were due to two independent genes, each with additive effects:

4 ft ♀ × 6 ft ♂
(aabb) ↓ (AABB)

all 5 ft
(AaBb)

5 ft ♀ × 5 ft ♂
(AaBb) ↓ (AaBb)

$^1/_{16}$ 4 ft:$^4/_{16}$ 4.5 ft:$^6/_{16}$ 5 ft:$^4/_{16}$ 5.5 ft:$^1/_{16}$ 6 ft
(aabb) (aaBb / Aabb) (AaBb / aaBB / AAbb) (AABb / AaBB) (AABB)

4 5 6

Figure 14–2 Monogenic vs. multigenic inheritance.

C. If height were due to many genes, each with additive effects:

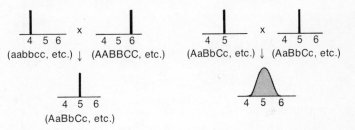

4 5 6   x   4 5 6
(aabbcc, etc.) ↓ (AABBCC, etc.)

(AaBbCc, etc.) x (AaBbCc, etc.)

4 5 6
(AaBbCc, etc.)

D. If height were due to many genes plus some environmental effects:

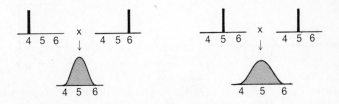

4 5 6   x   4 5 6

4 5 6

4 5 6   x   4 5 6

4 5 6

Quantitative traits are called polygenic traits because it is possible to show, at least in some cases, that many genes are involved. For height, concordance rates on MZ twins are higher than on DZ twins, suggesting a genetic input. But this does not tell us how many genes affect height. All we can say is that the bell-shaped distribution curve of heights in the population is consistent with the involvement of many nondominant (additive) genes and with diverse enviromental effects. We also know that environmental differences concerning height are important because people in industrial societies are significantly taller than they were 50 to 100 years ago (Fig. 14–3).

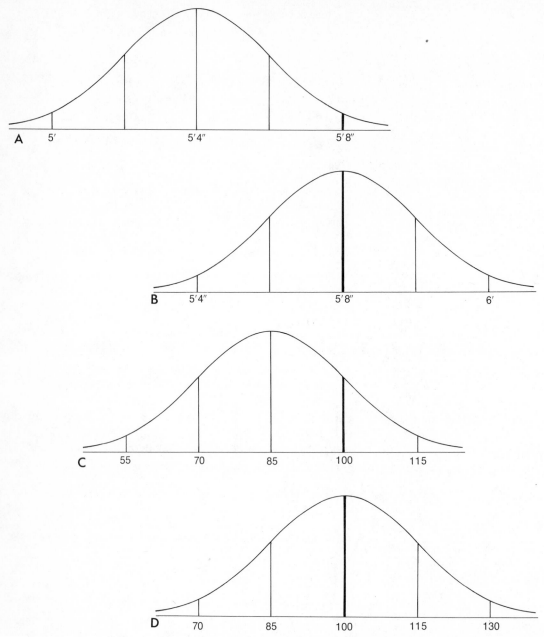

Figure 14–3 Height vs. IQ. A, Height distribution of European men four generations ago. B, Height distribution of European men now. C, IQ distribution of U.S. blacks now. D, IQ distribution of U.S. whites now.

I.Q. and Intelligence

We all form an impression of the intelligence of our friends and relatives. We think that some of them are brighter than others. But few of us know their I.Q. scores. If we were told their scores we would probably encounter some surprises — because many of us make the tacit assumption that I.Q. scores measure intelligent behavior or "general" intelligence.

Intelligence quotient (I.Q.) is defined as a function of "mental age" divided by chronological age. One's I.Q. is 100 if one has an average I.Q. test score for one's age. The hidden assumptions in this definition are that the test corresponds fairly to what an average person in a homogeneous cultural setting would know, that the test is free of socioeconomic bias, and many others. I.Q. scores have been shown to be correlated with grades obtained in school, socioeconomic status, educational level of one's parents, and other factors as well. Some people in the past have considered the basis of one's I.Q. score to be strictly genetic, while others have held rigidly environmentalist views. Today most people agree that both heredity and environment play a role in I.Q. Many would go further and say that I.Q. tests have caused more problems than they have solved.

The first I.Q. test was developed in France for the express purpose of identifying able students and affording them an opportunity to realize their potential for scholarship. I.Q. tests were soon adapted for use in the U.S. and were first used on a massive scale on recruits in World War I. The test results on the recruits and on recent immigrants showed that those of northern European ancestry scored higher than those from southern and eastern Europe. This finding parallels the European emigration pattern into the U.S., with people from southern and eastern Europe generally arriving later than those from farther north. Since few southern European immigrants spoke English, clearly they and their offspring in the U.S. would have less success in school than the northerners and would do poorly on I.Q. tests given in English on that account alone. But southern European immigrants also had cultural systems that differed from those of northerners, another important factor in their relative failure on the tests. However, such explanations were not fashionable at the time. The test results were accepted by the intellectual and political elite (largely of northern European stock) as showing that southern Europeans were genetically inferior to northern Europeans (and thus had less personal worth?).

I.Q. tests now are widely used in U.S. public schools and constitute a major determinant in evaluating a child as brilliant, merely educable, or retarded. (But a recent court decision in California forbids as discriminatory the use of I.Q. tests to determine "tracking" in schools.) And because blacks, Chicanos, and other minority groups today score lower on the average than do whites on I.Q. tests, the tests are still lauded by some whites as proving the genetic inferiority of minorities. But today a much larger group of citizens condemns I.Q. testing as an excuse for the continuation of racial prejudice.

Furthermore, because low-income persons have on the average lower I.Q. scores than those of higher-income persons, poor people are regarded by some groups as genetically inferior. In addition, early research seemed to show that low-income families had more children than those with high incomes, giving rise to fears in the "establishment" that poor, stupid people were out-reproducing all others and would thus bring about a decrease in the intelligence of the population. Fortunately, these fears can now be laid to rest: a variety of solid data shows unequivocally that when unmarried and otherwise childless adults are taken into account, most of them fall in the low I.Q. group (many low I.Q. people are institutionalized and thus were ignored by early investigators) (See Fig. 14–4). Thus there is actually strong evidence that persons with I.Q. scores over 100 are out-reproducing those with lower I.Q.'s. This finding may be of considerable comfort to some people, but it has meaning only if I.Q. measures something akin to intelligent behavior, which, although undefinable, is something everyone suspects the human species can use more of. Another important point here is that artistic and creative ability, which we could also use more of, are in no way measured by I.Q. tests.

Simply because of the social and political significance of I.Q. scores, we will now consider briefly what is known about their genetic-environmental basis. I.Q. scores are more highly correlated for MZ twins than for DZ twins. Interestingly, how-

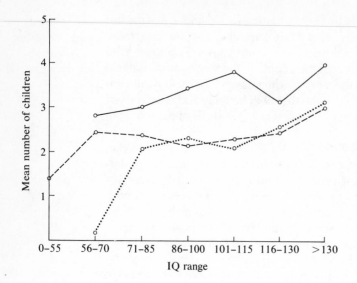

Figure 14–4 Differential fertility and intelligence as measured in three different studies. (After McClearn, G. E., and DeFries, J. C.: Introduction to Behavioral Genetics. San Francisco, W. H. Freeman, 1973.)

ever, the correlations between test scores of single individuals at different ages during childhood are much lower than either the MZ or DZ correlations. Between 2.5 and 17 years the average middle-class child's I.Q. changes over a range of 28 points (McCall et al., 1973). It seems that repeated testing hits "spurts and lags" at different points in development. Be this as it may, Jinks and Fuller (1970) carefully analyzed studies of the genetic basis of I.Q. scores and concluded that at least 20, but perhaps 100, genes are involved in the phenotypes of this complex trait. But whether individual differences in I.Q. score are due to differences among 20 to 100 gene sites, to differences in home environments, or both, is not known.

In the U.S. white population the average I.Q. score is 100; in the U.S. black population it is 85 (Fig. 14–3). The range of scores is the same in the two populations, from near zero to over 150. Are the differences in I.Q. score within the white population (up to 150 points) due to the same causes as the difference in average score between the white and black populations (15 points)? No one knows. Suppose all U.S. whites live in an optimal environment for maximizing their I.Q. score; all differences in I.Q. within the white population are then due solely to genes. Suppose further that all U.S. blacks live in a uniform but suboptimal environment; again, all the I.Q. differences within the black population would be solely due to genes. Then the shape of the I.Q. distribution curve for blacks and whites would be very similar, but the average score for blacks would be lower than that for whites. This is essentially what is found on actual testing. But is the explanation given above the true one?

In a recent book, Jensen (1973) has extensively considered various environmental factors that might contribute to the 15-point difference in average I.Q. score between blacks and whites. He has concluded that these factors can account for only 25 to 50 per cent of the difference between the two groups. The factors he discusses include inequality of schooling, motivation, cultural bias in I.Q. tests, sensory-motor differences, language deprivation, and nutrition. He concludes, "All the major facts would seem to be comprehended quite well by the hypothesis that something between one-half and three-fourths of the average I.Q. difference between American Negroes and whites is attributable to genetic factors, and the remainder to environmental factors and their interaction with genetic differences" (p. 363).

Many investigators think Jensen and others have woefully underestimated the importance of genetic-environmental interaction (e.g., the interaction between mother and infant who share genes as well as an environment). A striking study by Zajonc (1976) suggests that the child-spacing interval in families, which differs between blacks and whites, makes a large difference in the intellectual development of children.

A recent report by Willerman and his associates (1974) counters Jensen's argument that there is only a minor influence of environment, or genetic-environment interactions, in group differences in I.Q. scores. These investigators found that at four years of age black-white interracial children with white mothers had I.Q. scores nine points higher (102 vs. 93) than those with black mothers. Since there was no difference in birth weight, length of pregnancy, or other measurable variables between the two subgroups studied, the investigators interpret the results to mean that postnatal factors contribute substantially to I.Q. scores. (The white and black mothers, however, did not differ significantly in socioeconomic status or educational level.)

Heber and co-workers have shown that black infants placed in enriched environments five days per week for five years have average I.Q. scores 30 points higher than expected (125 rather than 95). Such results (and those of similar programs where the parent is involved directly in improving the child's development) invite skepticism about genetic variability as the major input in I.Q. differences within the U.S. black population—that is, that I.Q. scores are highly heritable in blacks. Although studies on the heritability of I.Q. scores in the *white* population give a value somewhere between 50 and 80 per cent, there are no data on heritability of I.Q. scores in blacks; moreover, heritability estimates for whites tell us nothing about the cause of the average difference in scores between whites and blacks—whether it is due mainly to hereditary or environmental differences. Furthermore, Kamin (1974) has shown that many of the heritability studies on whites were poorly designed and that their results misleadingly suggested that genetic differences produced most of the variability in I.Q. scores found within the white population.

MAN AND BEAST

A number of well known books on *human* social behavior, stressing aggression and competition and understanding cooperation, are entertaining but lack logical consistency and comprehensiveness. In contrast, books by George Schaller, Jane Goodall, and others on *animal* social behavior observed under natural conditions are generally first-rate scientific reporting. Nevertheless, the general tendency has been to oversimplify and overgeneralize from observations on animal behavior in the wild to assertions about human behavior, usually by omitting the importance of learning in the behavior of both humans and animals. Moreover, animals are not just imperfect approximations of humans. A comparison of different animals reveals more adaptive diversification to different ecological situations than it does a progressive development of organisms with human-like nervous systems.

Enormous evidence points to the evolution of humans from ape-like ancestors (Chap. Nineteen); yet this does not mean, as some science popularizers have stated, that our behavioral patterns in sex, rage, and love are ape-like. Human behavior is a product of genetic predisposition (or limitation), cultural norms, past history of the individual, and present circumstances. Is our behavior then not innate, instinctive, inevitable? Neither our genes nor the "devil" make us behave badly or well. Most of our behaviors, besides simple reflexes, are learned ones, but whether learned well or not depends on our motivation, opportunity, state of health, age, and other factors. The fact that these behaviors are based on both genic and cultural evolution over several million years of human distinctiveness from the apes is of enormous importance, but we do not have access to "fossil behaviors" to show what changes have occurred and why. If we observe the behaviors of contemporary primates in the wild, we find them to be very different from those of humans. If we look at hunting-gathering societies in existence now (Australian aborigines, African Pygmies), we find them to be very different from industrial societies. Cooperation and competition in these different contexts are much too labile in their meanings to allow us to develop rigorous theories of what makes us do what we do.

We have only bits and pieces of solid information about differences between human groups in terms of innate propensities. For example, Freedman and Freedman (1969) have tested behavioral differences between newborn Chinese and white infants in a hospital nursery (in the U.S.). The two groups differed mainly in temperament, the Chinese infants being somewhat more imperturbable, more readily accepting of external change, and more readily self-consolable (not crying as long if they were not picked up). But in sensory development, maturity of the nervous system, motor development, and social responsivity the two groups were alike.

Human behavior varies over a wide range. Overall, it is the result of successful attempts to cope with the physical, social, and personal environment over thousands of generations—no more, no less.

SUMMARY

The bases of human behavior, both individual and social, are genetic, developmental, neurological, and environmental. Most behaviors of interest are not traceable to single-gene effects in any clear-cut way. Many behaviors seem only remotely connected with genes. However, schizophrenia and manic-depression, two severe mental disorders, are currently being scrutinized for a biochemical and genetic basis because the clues available seem to point in that direction. The behavioral outcome of the known genetic abnormality XYY is also under consideration. The deluge of reports on human behavioral traits reaches its high tide when we consider the normal range of I.Q. scores in populations. The genetic basis of I.Q. is even less clear than that for abnormal traits like the psychoses. The human species is behaviorally unique in many ways because of its unique evolutionary past, a past molded by natural selection.

> The song of men all sorts and kinds,
> As many tempos, moods and minds,
> As leaves are on a tree . . .
>
> —Ralph Hodgson, "The Song of Honor" (1917)

STUDY QUESTIONS

1. Manic-depression may be a dominant X-linked illness. If this is the case,
   (a) mothers and daughters of male manic-depressives will be affected more often than the fathers and sons.
   (b) female manic-depressives will have as many mothers affected as fathers affected.
   (c) sons and daughters of female manic-depressives will be equally affected.
   (d) all of the above
   (e) none of the above

2. Suppose that for quantitative trait Q, female monozygotic twin pairs are more different (more often discordant) than are male MZ pairs. How might this difference between the sexes be accounted for?

3. White prisoners have been found to be 2.5 times more likely to be XYY than black prisoners in the U.S. XYY newborns occur once in 1000 male births among whites and perhaps no more than once in 5000 male births among blacks (scanty data). Among the factors that may be involved in the white-black karyotype difference in prisoners are the following: (a) blacks have a higher death rate than whites in the U.S.; (b) blacks are generally poorer than whites in the U.S.; and (c) poor people are imprisoned more often than others are. Discuss the probable cause of the high rate of XYY's among white prisoners.

4. Potato plants grown from different "eyes" of a single potato constitute a genetically uniform clone. Members of such a clone grown in diverse environments have a bell-shaped distribution of heights (as in Fig. 14–3, C, F_1 on right side). But if the clone is grown in a constant environment it has a skyscraper distribution (Fig. 14–3, C, F_1 on left side). How does this finding relate to the genetics of height in potato plants?

5. Comment on the following statement by Rene Dubos: "We cannot escape from the zoos we have created for ourselves and return to wilderness, but we can improve our societies and make them better suited to our unchangeable biological nature" (Montagu, 1973, p. 91).

6. Key Terms:

behavior	genetic heterogeneity	heritability
pleiotropy	monozygotic	I.Q.
polygenic	dizygotic	clone
schizophrenia	concordance	XYY
manic-depression	discordance	quantitative trait

REFERENCES*

Beckwith, J., D. Elseviers, L. Gorini, C. Mandansky, and L. Csonka. Harvard XYY study. Science 187:298, 1975.

Boraonkar, D. S., and S. A. Shah. The XYY male—or syndrome? Progress in Medical Genetics 10:135–222, 1974.

Freedman, D. G., and N. C. Freedman. Behavioral differences between Chinese-American and European-American newborns. Nature 224:1227, 1969.

Heber, R., et al. See pp. 159–163 in Loehlin et al. cited below.

Heston, L. L. Psychiatric disorders in foster home reared children of schizophrenic mothers. British Journal of Psychiatry 112:819–825, 1966.

Heston, L. L. The genetics of schizophrenia and schizoid disease. Science 167:249–256, 1970.

Hook, E. B. Behavioral implications of the human XYY genotype. Science 179:139–150, 1973.

Inouye, E. Some considerations in the methodology of human behavior genetics. Social Biology 20:241–245, 1973.

Jensen, A. R. Educability and Group Differences. New York: Harper & Row Publishers, Inc., 1973.

Jinks, J. L., and D. W. Fuller. Comparison of the biometrical, genetical, Mava, and classical approach to the analysis of human behavior. Psychological Bulletin 73:311–349, 1970.

---

*Note: items cited in the References are also recommended for suggested reading.

Kamin, L. J. The Science and Politics of I. Q. New York: Halsted Press, Division of John Wiley & Sons, Inc., 1974.

McCall, R. B., M. I. Appelbaum, and P. S. Hogarty. Developmental changes in mental performance. Monographs of the Society for Research in Child Development 38:1–84, 1973.

Mage, M. Harvard XYY study. Science 187:299, 1975.

Mendlewicz, J., and J. D. Rainer. Morbidity risk and genetic transmission in manic-depressive illness. American Journal of Human Genetics 26:692–701, 1974.

Montagu, A. (ed.). Man and Aggression. Ed. 2. New York: Oxford University Press, 1973.

Shaskan, E. G., and R. E. Becker. Platelet monoamine oxidase in schizophrenics. Nature 253:659–660, 1975.

Stern, C. Principles of Human Genetics. Ed. 3. San Francisco: W. H. Freeman & Co., 1973. See Chapters 27 and 31.

Willerman, L., A. F. Naylor, and N. C. Myrianthopoulos. Intellectual development of children from interracial matings: performance in infancy and at four years. Behavior Genetics 4:83–90, 1974.

Wyatt, R. J., M. A. Schwartz, E. Erdelyi, and J. D. Barchas. Dopamine beta-hydrolase activity in brains of chronic schizophrenic patients. Science 187:368–370, 1975.

Zajonc, R. B. Family configuration and intelligence. Science 192:227–236, 1976.

SUGGESTED READINGS

Akiskal, H. S., and W. T. McKinney. Depressive disorders: toward a unified hypothesis. Science 182:20–29, 1973.

Brace, C. L., and F. B. Livingstone. On creeping Jensenism. In Race and Intelligence, C. L. Brace, G. R. Gamble, and J. T. Bond (eds.). Washington, D.C.: American Anthropological Association, 1971.

Cronbach, L. J. Five decades of public controversy over mental testing. American Psychologist 30:1–14, 1975.

Culliton, B. J. Patients' rights: Harvard is site of battle over X and Y chromosomes. Science 186:715–717, 1974.

Culliton, B. J. XYY: Harvard researcher under fire stops newborn screening. Science 188: 1284–1285, 1975.

Ehrman, L., and P. Parsons. The Genetics of Behavior. Sunderland, Mass.: Sinauer Associates, Inc., 1976.

Elston, R. C. Methodologies in human behavior genetics. Social Biology 20:276–279, 1973.

Feldman, M. W., and R. C. Lewontin. The heritability hang-up. Science 190:1163–1168, 1975.

Goldsby, R. A. Race and Races. New York: Macmillan Publishing Co., Inc., 1971.

Hook, E. B. Racial differences in the prevalence rates of males with sex chromosome abnormalities (XXY, XYY) in security settings in the United States. American Journal of Human Genetics 26:504–511, 1974.

Kety, S. S., D. Rosenthal, P. H. Wender, F. Schulsinger, and B. Jacobsen. Mental illness in the biological and adoptive families of adopted individuals who have become schizophrenic. In Genetic Research in Psychiatry, R. Fieve, D. Rosenthal, and H. Brill (eds.). Baltimore: Johns Hopkins Press, 1975, pp. 147–165.

Loehlin, J. C., G. Lindzey, and J. N. Spuhler. Race Differences in Intelligence. San Francisco: W. H. Freeman & Co., 1975.
    A summary statement marred by internal contradictions; see review by R. C. Lewontin, American Journal of Human Genetics 28:92–97, 1976.

McClearn, G. E., and J. C. DeFries. Introduction to Behavioral Genetics. San Francisco: W. H. Freeman & Co., 1973.

Miner, G. D. The evidence for genetic components in the neuroses: a review. Archives of General Psychiatry 29:111–118, 1975.

Money, J., C. Annecillo, B. VanOrman, and D. S. Borgaonkar. Cytogenetics, hormones, and behavioral disability: comparison of XYY and XXY syndromes. Clinical Genetics 6: 370–382, 1974.

Murphy, J. M. Psychiatric labeling in cross-cultural perspective. Science 191:1019–1028, 1976.

Rosenthal, D. Genetic Theory and Abnormal Behavior. New York: McGraw-Hill Book Co., 1970.

Samuel, D., and Z. Gottesfeld. Lithium, manic-depression, and the chemistry of the brain. Endeavor *32*:122–128, 1973.

Singh, M. M., and S. R. Kay. Wheat gluten as a pathogenic factor in schizophrenia. Science *191*:401–402, 1976.

Sutton, H. E. An Introduction to Human Genetics. Ed. 2. New York: Holt, Rinehart & Winston, Inc., 1975.

Winick, M., K. K. Meyer, and R. C. Harris. Malnutrition and environmental enrichment by early adoption. Science *190*:1173–1176, 1975.

Witkin, H. A., et al. Criminality in XYY and XXY men. Science, *190*:547–555, 1976.

Yurchevco, H. A Mighty Hard Road: The Woody Guthrie Story. New York: McGraw-Hill Book Co., 1970.

Motherhood. Micronesian carvings in the collection of J. Royce.

Chapter Fifteen

GENETIC COUNSELING

Genetic counseling is the long-term process of informing people of the risk that a family member will inherit a genetic defect, of the means available to minimize such risks, of the treatment available for the defect, and related concerns. Genetic counselors inform people of such facts as the developmental fate of a child with a certain disease and the risk of recurrence of the disease in future children. They do not tell people what to do about the risk information — whether to avoid having future children, to undergo sterilization, or to terminate a pregnancy. Genetic defects can crop up in any family; a good estimate is that about three per cent of the population would benefit by receiving genetic counseling, according to J. A. F. Roberts.

WHO COUNSELS AND WHERE

There is a broad range of medical-genetic services available in most parts of the U.S. Different services are appropriate for different kinds of genetic problems. In general, genetic counseling is most effectively done by several professionals working together in a clinic. In a large clinic a medical doctor with some training in genetics diagnoses and treats the inherited ailment, while a geneticist with some medical training collects the family medical history as far back as possible, constructs a family pedigree, and advises the family of the risk of further affliction in various branches of the family. In a small clinic a single medical geneticist may perform all these services. Public health nurses, medical social workers, and others can help an afflicted family with its economic problems, with medical follow-ups, and with psychological problems brought on by the illness. Other professionals in the clinic, or in cooperating laboratories that may be hundreds of miles away, do the specialized work of growing cells from the patient in culture flasks, performing chemical tests on the cultured cells for possible gene malfunction, or analyzing the cells for their chromosome number and pattern. If a woman is already pregnant with a possibly defective fetus, amniocentesis may be performed between the twelfth and sixteenth week after conception to obtain fetal cells for testing for certain chromosomal or biochemical abnormalities. The option of therapeutic abortion is made available following diagnosis of abnormality in utero.

Often the genetic counseling clinic workers are so mutually interdependent that the medical doctor cannot make a diagnosis until he receives the results of chemical-genetic tests: he needs to know the chromosome number and pattern before he diagnoses Down's or certain other syndromes. He needs to know the enzyme level determined in laboratory tests for a diagnosis of Tay-Sachs disease (see Box on TSD, Chap. Ten). He needs to know the electrophoretic pattern of red blood cell proteins before diagnosing sickle-cell anemia or a thalassemia condition (Chap. Ten). Likewise, the geneticist who plots the probabilities that individuals in the family will pass on an afflicting gene also needs to know these test results. Often on the basis of his studies, the geneticist can alert the family pediatrician or other medical workers that members of the family who are not serviced by the local clinic, and who may, in fact, live in a distant city, are at risk for a certain defect.

No medical doctor can be expected to be conversant with the 1500 or more genetic defects now recognizable. Most of them are so rare that he or she will never see a single case of them in a lifetime of medical practice. But a doctor alerted to the chance that a certain genetic defect may occur in a patient—as a result of Mendelian segregation of alleles—is then primed to diagnose the ailment and begin treatment promptly, as is often critical for success in alleviating diseases like PKU and pyloric stenosis. Babies with PKU must be put on a low phenylalanine diet very soon after birth and kept on it for at least six years to prevent long-term brain damage (Chap. Ten). Pyloric stenosis babies have a thickened muscle constricting the lower outlet of the stomach and thus require simple surgery a few weeks after birth. Otherwise the child may die from vomiting and dehydration. Fortunately, however, heredity plays a small role in this disease; the risk of recurrence in a family is at most ten per cent.

Many medical schools in the U.S. now have genetic counseling clinics associated with their affiliated hospitals. Many other hospitals have or are planning to have a genetic counselor on the staff. In addition, some family doctors, including obstetricians and pediatricians, provide genetic information at times. But there are several problems with the family doctor approach. First, many older doctors have received no education in genetics (its importance to biomedicine has only lately been recognized). Secondly, they therefore give out misinformation to their patients frequently enough to be alarming. Thirdly, doctors have insufficient time to work out a detailed family pedigree, determine the genetic implications for each individual in the pedigree, and determine if the family members sufficiently understand the genetic risk to make appropriate decisions about future childbearing. Finally, doctors may be paternalistic toward their patients to the extent that they may decide unilaterally on sterilization or abortion for individuals in situations when the genetic facts do not even warrant it (Fraser, 1974). Thus, despite reports that patients feel more comfortable with their family doctor than they do with strangers in a specialized clinic, all indications are that unless their doctor is soundly educated in genetics, they will be better off consulting a genetic counselor about inherited diseases in order to maximize their chances of getting accurate and complete information for family planning. In the many instances when there is no family doctor, the genetic clinic approach is the only one feasible in any event.

WHO RECEIVES COUNSELING

Most inherited diseases are rare recessive traits. They range in frequency from about 1 in every 100,000 births to about 1 in 500 births (as is the case for sickle-cell anemia in black populations, where about eight per cent are carriers for this autosomal recessive disease). In contrast, chromosomal abnormalities collectively affect 1 in every 150 to 200 births; many of these show severe birth defects. Except in a few cases it is not yet possible, let alone practicable, to test people to find out which of them are carriers for genetic defects.

In the U.S. the only mass screening for carriers so far has been for Tay-Sachs (among Jews of European origin) and for sickle-cell anemia (among U.S. blacks). The current test for cystic fibrosis (CF) carriers (who constitute about five per cent of the white population) is unreliable in that it produces some "false negatives"; that is, some people already known to have the CF gene appear not have it according to their test results. For about 20 other recessive diseases it is now possible to identify carriers, but either it is too expensive yet to do it on a mass-screening basis or it is almost futile because the disease is so rare (i.e., so thinly spread throughout a large population). Thus prospective genetic counseling for most recessive diseases based on the identification of carriers by mass screening is still at the talk stage. But it may come soon.

In the meantime, although most people are carriers for three to eight recessive genetic defects, they have no way of knowing it—unless they marry a person who carries the same defect; then the knowledge may unfortunately be thrust upon them by their having a defective child. Such people have lost out on the genetic gamble we all take in deciding to reproduce. It is not surprising, then, that statistics show that 90 per cent of couples who present themselves at genetic counseling clinics are healthy people who already have one defective child. The rest of the couples who ask for counseling either are related to each other and thus fear some genetic defect in their children due to inbreeding—which again would require a recessive trait—or are afraid of having a defective child because some family member, such as an uncle or grandparent, already has some defect. Their family defect may be autosomal dominant or recessive, X-linked, or not inherited at all. If there is an autosomal dominant gene in the family causing the defect, then in most cases it will show up every generation, say from grandmother to daughter to some of the grandchildren. If autosomal recessive, it may have appeared only once in the family, unless there are cousin marriages or other close inbreeding ties. If there is an X-linked disease, it is almost always recessive and shows up sporadically in males in the family tree (as do colorblindness and hemophilia).

Genetic Risk

In the genetic clinic, after biochemical or chromosomal tests on a couple, and after the construction of a pedigree of all afflicted relatives in the family (including even remote cousins if such information can be obtained), the couple is informed of the risks of their having a child with a genetic problem, or having a second such child. Accurate diagnosis of the afflicted child and relatives is essential for assessing genetic risks. Diagnosis often requires the counselor's verification by obtaining past hospital or other medical records. Knowledge of the genetic ratio for the particular trait in question is also necessary to establish future risks of the disease in the family. For the many genetic diseases that are caused by segregating recessive alleles of a single gene, the establishment of risk figures is usually simple. For example, the chances are 1 in 4 overall when both parents are carriers (autosomal), or 1 in 2 for boys only for X-linkage when the mother is a carrier. Some clinical diagnoses, however, break down into two or more genetic categories when the family pedigree is analyzed. Childhood muscular dystrophy, for example, comes in two forms—X-linked and autosomal recessive—and the risks of recurrence thus differ in the two cases. Such genetic heterogeneity is a fairly common complication in the causation of genetic defect (Chap. Fourteen).

But future risks of genetic disease are not calculated in all cases simply by knowing Mendelian segregation ratios. Conditions caused by polygenic inheritance (Chap. Fourteen) or by new mutations pose difficult problems. Polygenic or multigenic inheritance seems to be the reason for the recurrence in certain families of diseases like diabetes mellitus, pyloric stenosis, and spina bifida (exposed spinal cord). Since an unknown number of genes seem to act together, cumulatively, to produce such diseases, we cannot separate out for study each gene involved and establish its segregation pattern to the offspring. All that can be done in these cases, at least with the present foggy knowledge, is to judge the risk of the disease recurring in the family on the basis of studies of previous cases. This is called estimating the risk empirically.

For example, C. O. Carter has shown from studies on a number of families in Britain that a normal couple who already have one child with cleft lip and are worried about having a second such child have a 4% (1 in 25) risk of having a second debilitated child. This is not a serious risk overall, since "1 in 40 of all white children at birth have some serious congenital malformation" due to genes or other causes (Roberts, 1970). However, in planning their families, people are more often swayed by the emotional and economic burden of an affected child than by stark risk figures.

New mutations, although rare, must be considered by the genetic counselor, especially for dominant diseases. In cases where there is one child in the family with a dominant defect such as achondroplastic dwarfism (see p. 22.), but neither parent or any other relative shows the defect, it may be that a new mutation has occurred (giving a risk of recurrence of almost zero). Alternatively, a mutant gene may be present but not expressed in one of the parents that acts as a recessive in some cases and a dominant in others which is at fault. The expression of such a gene may depend upon the life circumstances of the individuals involved. An infectious disease, poor nutrition, or other factors may bring out the gene's effect in one person while another person may escape such a precipitating factor. These genes that act with variable manifestation in different individuals (variable penetrance) can be handled by genetic counselors only on the basis of empiric risks, just as with polygenic traits. In any event, the empiric risk of recurrence of a variably penetrant defect in a family is generally low.

Down's syndrome is mostly due to new mutations at the chromosomal level: an error in chromosome segregation during cell division produces one daughter gamete with an extra chromosome (No. 21) which, when combined with a normal gamete, forms an abnormal zygote (24 + 23 = 47 chromosomes). Thus most Down's syndrome children occur sporadically in families. That is, all parents have a low risk, 1 in 700 births, or 0.14%, of such a mutation occurring in one of their reproductive cells. (But, as discussed in Chap. Six, older women have a higher risk and younger ones a lower one.)

In talking about risks of genetic disease it is important to remember that any couple runs the risk of about 2 in 100 of having a child with *some* serious genetic problem. That is, for every one hundred babies born, on the average two will have a genetic defect. This is called the random risk (Roberts, 1970). Far above the random risk of 2% is the high risk group of genetic defects, with risks to offspring ranging from 1 in 10 (10%) to 1 in 2 (50%). The medium and low risk groups of genetic defects, to which people react in less shocked fashion than they do to the high risk group, range from about 5 to 10% for the recurrence of a defect in the family. Families faced with risks for genetic defects of greater than 10% often have no more children on that basis.

High risk conditions include simple genic dominants (50% risk with one parent affected), simple recessives (25% risk with carrier parents), and also certain chromosomal anomalies. A parent with a D/G translocation and one Down's syndrome child has at most a 10% empiric risk of having a later child with Down's. Why the translocation type of Down's anomaly repeats only that often is not known, but many embryos with the Down's chromosome pattern abort spontaneously and only a minority survive to birth or beyond. In cases where a mother or father has been shown to have a translocation which can produce Down's syndrome, amniocentesis and chromosome analysis of fetal cells now make this previously high risk inherited form of Down's a very low risk situation. Parents now have the option to abort a fetus if amniocentesis shows it to have a Down's karyotype. Since most children of such parents, however, will be normal, amniocentesis screening in most cases proves to be negative for an affected child and thus permits the parents to go ahead with the pregnancy without anxiety.

The risk of retinoblastoma (an eye cancer) for a child with one parent who has the disease is 50% because the condition is dominant; yet the recurrence risk for normal parents with one child with retinoblastoma in both eyes is low, less than 10% empirically (Carter, 1969). These rare cases of individuals with normal parents probably include some with new mutations. (Such cases, if they could be separated out, would have an essentially zero risk of recurrence.) And they may include, as well, cases where the gene is poorly penetrant or where the new mutation occurred not in a single reproductive cell of the parent but in a stem cell of parental reproductive tissue, causing more than one sex cell to be affected.

Once a case of a recessive or poorly penetrant dominant disease is found in a family, it is sometimes possible to test family members to find out if they are carriers. For a recessive disease, it is indirectly known that the parents are indeed carriers once an affected child is born. But other adults in the family who have not reproduced yet may also be carriers, as may phenotypically normal siblings of the affected child. With a disease called variegate porphyria, for example, family members who are medically normal can be shown to be heterozygotes if they show increased levels of the chemicals known as protoporphyrins in their stools. As this defect is rare, however, two heterozygotes are very unlikely to meet and marry: only by marrying a relative is a heterozygote venturing into genetic shoals, and even then the relative can now be tested for heterozygosity for this defect before children are contemplated.

While there are high risk genetic diseases of a specific nature, it is important to note that some conditions with multiple causes such as blindness and deafness are largely genetic in origin. By 1950 genetic causes were shown to account for 68 per cent of cases of blindness in schoolchildren in Britain, and by 1964 they accounted for half of the cases of complete deafness in Britain, according to surveys conducted by N. C. Nevin. These high frequencies from genetic causes are thought to be due to the decline in blindness and deafness from infections and other environmental factors.

WHAT COUNSELEES LEARN

So far we have been assuming that genetic counseling patients, or their parents, understand all the medical and genetic information presented to them. Such is not the case, according to numerous studies. Emotional conflicts are the main barrier to understanding. Many people entering a genetic counseling clinic, for whatever reason, bring with them severe fears of a genetic "taint" which will cause all their children to be diseased. They are sometimes also beset with genetic mis-information of other kinds. Some patients are in such a state of shock when they first find out they carry an inherited defect that they don't even hear the balance of the counselor's message. A few patients find their defective gene so emotionally unacceptable that they later deny they had ever received genetic counseling. They may also deny the disease is genetic when it definitely is. This denial means that they cannot then see the point of using effective contraception to avoid having a later child with the same defect. In a study by C. O. Leonard and others (1972) 34 per cent of families with genetic disease were not using effective contraception after the birth of a defective child. Fathers of an affected child have even been known to deny paternity or to initiate divorce proceedings because they were unwilling to accept their genetic defect.

Some patients who receive genetic counseling information cannot use it for making an objective decision about childbearing because of their faulty under-standing of statistics. One woman thought that since she was not planning to have a family of ten children, the genetic risk figure of 1 in 10 told to her did not apply to her at all. Moreover, because a genetic counseling clinic may be treating a number of patients with the same disease on the same day, parents sometimes judge that the recurrence risk figure given to them by the counselor is unrealistically low, Leonard reports. Other patients think they are just naturally unlucky and are bound to have defective children, no matter what the odds in genetical terms. This gloomy outlook is reinforced if they have already had a run of bad genetic luck. A run of good luck genetically has been known to have the opposite psychological effect.

Another reason for imperfect reception of medical-genetic counseling informa-tion is lack of understanding of the biological facts of human reproduction. Talking about genes, proteins, and cells to adults who have never been taught human biology in school is not very successful in a genetic counseling context. Six months or more after being counseled, only about half the families with genetic disease interviewed by Leonard recalled enough of the medical-genetic information to apply it to making a rational decision about having more children.

On the other side of the coin, genetic counselees often ask questions about "their" family disease for which no answers currently exist ("How long will my Down's syndrome child live?" etc.). Much research is still needed to obtain reliable information on the risks associated with numerous genetic diseases.

ECONOMICS OF GENETIC COUNSELING

Genetic diseases affect a startling number of families. In a large hospital 17 per cent of pediatric patients had a primary diagnosis of a genetic disorder (Day and Holmes, 1973). The recognition of a genetic involvement in medical problems increases daily, so the 17 per cent figure represents only the crest of the wave. To underscore the magnitude of the problem, consider Down's syndrome. It occurs in about 1 in 700 births. Etzioni (1974) estimates that the Down's children in the U.S. cost the afflicted families and the government about $1.7 billion per year until 1974, when amniocentesis and therapeutic abortion for Down's fetuses became widespread options. If amniocentesis is used prospectively in the U.S. on all older pregnant women, abortion of the cases of Down's syndrome fetuses detected could cut in half the incidence of Down's cases in twenty years. That is, half the antici-pated 57,000 to 83,000 Down's children in the U.S. during the next decade would not be born, enormous medical bills would be saved, human suffering would be lessened, and scarce medical resources (now used for Down's patients) would be available for others.

OTHER ASPECTS OF MEDICAL GENETICS

It certainly isn't necessary to consult a genetic counselor to determine your ABO blood type or to determine whether an Rh negative mother might have an Rh incompatible child; such matters are widespread problems well understood by most medical personnel. Yet there are other aspects of genetics that intrude on and confound the medical scene often enough to warrant mention here, even though they are not part of genetic counseling. The best known of these is the genetic incompatibility of tissues that so often causes rejection of surgically transplanted hearts and other body organs. The genetic incompatibility of such tissues is minimized by using "compatible donors" (often tissue from a sibling) in combination with drugs which suppress the immunological reaction that produces rejection of tissue grafts by the body.

The medical penchant for prescribing drugs likewise often ignores individual genetic variation. A G6PD-deficient man working in the malarial regions of the tropics will have a severe attack of anemia if he takes the anti-malarial drug primaquine. Heterozygotes who carry the porphyria gene get stomach pains when they take barbiturates, and lactose-intolerant people (many Asians and Africans) get cramps and diarrhea if they drink milk (Chap. Nineteen). Such problems are becoming commonly known among geneticists as the result of continuing research. The large number of such problems only confirms our rich genetic diversity.

Another aspect of medical genetics is mass screening for inherited diseases. Although screening is sometimes followed by genetic counseling, it is a separate phenomenon (see Box).

GENETIC SCREENING

Phenylketonuria (PKU) was the first genetic disease for which large populations were screened. Sickle-cell anemia was the second such disease. Now Tay-Sachs disease and other, even rarer, genetic problems have joined the list. Laws have been passed in almost all the United States mandating screening tests for one or more genetic diseases for the appropriate "target" sector of the population.

Problems have arisen in these mass screening programs for a variety of reasons:
1. Carriers of the disease cannot be identified by the test—only homozygotes are identified (e.g., PKU).
2. The kind of test used gives some false diagnoses (e.g., PKU).
3. Many of the state laws on genetic screening require mandatory testing of newborns but do not require information to be given to the parents, let alone require their consent for the test (e.g., PKU).
4. No treatment is available for the condition other than therapeutic abortion (e.g., Tay-Sachs and chromosomal abnormalities).

5. The disease can be detected after birth but not beforehand, thus disallowing the possibility of therapeutic abortion (e.g., sickle-cell anemia, beta-thalassemia).
6. The psychological problems which arise from informing unsuspecting people that there is something wrong with their heredity are often not taken into consideration by the well-intentioned screening teams.
7. Follow-up genetic counseling is often not available for discussing alternatives in reproductive planning of carriers identified by the screening process.

Thus, although mass screening programs were initiated to alleviate human suffering (which they do to some extent) and at the same time to allow the government to save money on medical expenses, geneticists, psychologists, and legal scholars are currently giving screening programs a "second look," emphasizing public education and voluntary participation in the programs as the best approaches in view of our imperfect knowledge of all the ramifications of these genetic tests.

SUMMARY AND CONCLUSIONS

More genetic components of diseases are becoming known through research. The stage has now been reached when genetic specialists must be on the medical scene for effective handling of genetic diseases. These medical-genetic counselors should advise patients, and parents, with not only great expertise but great patience and understanding, since the problem lies not only with afflicted individuals in the present generation but has implications for all future generations. The ethical problems of medical-genetic counseling in terms of public welfare will be discussed in a later chapter. As an introduction to that discussion, consider the following remarks by L. R. Kass (1972):

> Do we know what constitutes a deterioration or an improvement in the human gene pool? One might well argue that the crusaders against genetic deterioration are worried about the wrong genes. After all, how many architects of the Vietnam war have suffered from Down's syndrome? Who uses up more of our irreplaceable natural resources and who produces more pollution—the inmates of an institution for the retarded or the graduates of Harvard College? It is probably as indisputable as it is ignored that the world suffers more from the morally and spiritually defective than from the genetically defective. (Pp. 18–19.)

REFERENCES*

Carter, C. O. Genetic counseling. Lancet 1:1303–1305, 1969.
Day, N. and L. B. Holmes. The incidence of genetic disease in a university hospital population. American Journal of Human Genetics 25:237–246, 1973.
Etzioni, A. Genetic fix. New Scientist, January 17, 1974.
Fraser, F. C. Genetic counseling. American Journal of Human Genetics 26:636–659, 1974.
Kass, L. R. New beginnings to life. In The New Genetics and the Future of Man, M. Hamilton (ed.). Grand Rapids, Mich.: Wm. B. Eerdmans Publishing Co., 1972.

*Note: items cited in the References are also recommended for suggested reading.

Leonard, C. O., G. A. Chase, and B. Childs. Genetic counseling: a consumer's view. New England Journal of Medicine 287:433–439, 1972.

Nevin, N. C. Genetics and preventive medicine. Royal Society of Health Journal 89:281–285, 1969.

Roberts, J. A. F. Introduction to Medical Genetics. Ed. 5. London: Oxford University Press, 1970.

SUGGESTED READINGS

Bylinsky, G. What science can do about hereditary diseases. Fortune, September, 1974, pp. 148–160.

Culliton, B. J. Amniocentesis: HEW backs test for prenatal diagnosis of disease. Science 190: 537–540, 1975.

Culliton, B. J. Genetic screening. Science 189:119–120, 1975, and 191:926–929, 1976.

Danes, B. S. Genetic counseling. Medical World News 7:35–41, 1970.

Fletcher, J., R. O. Robin, and T. M. Powledge. Informed consent in genetic screening programs. In Ethical, Social, and Legal Dimensions of Screening for Human Disease, D. Bergsma et al. (eds.). Birth Defects Original Article Series, Vol. X, No. 6, 1974, pp. 137–144.

McKusick, V. Genetic counseling. American Journal of Human Genetics 27:240–242, 1975. Statement of the American Society of Human Geneticists.

Milunsky, A. The Prevention of Genetic Disease and Mental Retardation. Philadelphia: W. B. Saunders Co., 1975.

Milunsky, A., and G. J. Annas. Genetics and the Law. New York: Plenum Publishing Corp., 1976.

Milunsky, A., and P. Reilly. The "new" genetics: emerging medicolegal issues in the prenatal diagnosis of hereditary disorders. American Journal of Law and Medicine 1:71–88, 1975.

Pearn, J. H. Patients' subjective interpretation of risks offered in genetic counseling. Journal of Medical Genetics 10:129–134, 1973.

Powledge, T. M. Genetic screening vs. personal freedom. Science Digest, September, 1975, pp. 48–55.

Ramsey, P. Screening: an ethicist's view. In Ethical Issues in Human Genetics, B. Hilton et al. (eds.). New York: Plenum Publishing Corp., 1973.

Roberts, J. A. F. Congenital malformations. Postgraduate Medical Journal 44:148, 1968.

Sly, W. S. What is genetic counseling? In Contemporary Genetic Counseling: Birth Defects. Original Article Series 9:5–18, 1973.

Stevenson, A. C., B. C. Clare Davison, and M. W. Oakes. Genetic Counseling. Philadelphia: J. B. Lippincott Co., 1970.

WHO Expert Committee. Genetic Counseling. WHO Technical Report Series, No. 416, 1969.

It is more important that a proposition be interesting than that it be true.

Alfred North Whitehead

Chapter Sixteen

GENES, POPULATIONS, AND EVOLUTION

The word evolution is used in two ways. It refers both to the changes that have occurred in the history of life forms on earth, and to the processes which are proposed to explain these changes. When we say man has evolved we mean that the human species has changed over time. These changes are inferred by the differences between our present bone structure and the fossil remains of ancient humans and prehumans, as well as by the conspicuous differences between us and other living primates. In the case of man, we often tacitly assume that human evolution has meant a change for the better, or a change towards perfection. But when we say that birds, trees, and gorillas have evolved, we do not necessarily assume that the underlying changes have led to bird-perfection or gorilla-perfection. Nor do we think that birds or gorillas are tending in the direction of human-hood. They are obviously going their own ways. And, if we could relax our anthropomorphic attitude a moment, we might acknowledge that the human species is simply going its own way too. From this viewpoint the human way is not intrinsically better than the bird way or the gorilla way. But it is different.

Evolution occurs when the allele frequency in a population is changing. What causes the frequency of an allele to change? A number of forces which allow allele frequencies in a population to change have been identified: genetic drift, migration, and mutation and selection. We will consider the impact of these evolutionary forces on human populations as we go along. But first let us consider what must happen if there were no evolutionary changes.

A static condition of life on earth could occur only if the environment were unchanging and the populations of living species did not change genetically. The latter requirement for nonevolution rules out change in the proportions of alleles found in the populations. In shorthand fashion we can summarize this nonchange as $\Delta q = O$. Here q is the frequency of one allele of any gene; if p is the frequency of all the other alleles, then $p + q = 1$ ($p + q$ = all the alleles for the gene); and any change (Δ) in q entails a change in p. The change in q, written delta (Δ)q, is then one way of expressing evolutionary change; that is, evolution occurs when $\Delta q \neq 0$ (when the allele frequency is changing). And evolution does not occur when $\Delta q = 0$, when the allele frequency is not changing.

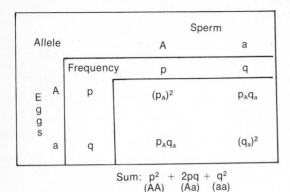

Figure 16–1 Multiplying allele frequencies in gametes to obtain genotypic frequencies.

HARDY-WEINBERG MODEL OF NONEVOLUTION

The formal way to describe $\Delta q = 0$ is called the Hardy-Weinberg genotypic equilibrium. It is a useful but oversimplified model which shows us how allele frequencies stay the same if evolutionary forces are inactive. By modifying the model to let one or more evolutionary forces become active, we can then visualize how evolution can come about.

The Hardy-Weinberg model, named after its two independent originators, says that given an infinitely large population where mating is "at random" (everyone having an equal chance of mating with everyone else), and given that $p + q = 1$, then $(p + q)_{females} \times (p + q)_{males}$ will give the frequencies of the genotypes in the population. (See Box on Population, below.) In the model these genotypic frequencies will be in equilibrium proportions and will not change from one generation to the next.

Let p be the frequency of allele A in a population and q be the frequency of the alternative allele a ($p_A + q_a = 1$). Since all the sperm and eggs in the population contain either A or a, if we randomly combine them, $(p_A + q_a)_{eggs} \times (p_A + q_a)_{sperm}$, we thus obtain the next generation of individuals:

$$p_A p_A + p_A q_a + q_a p_A + q_a q_a$$

WHAT IS A POPULATION?

A population is a group within a species sharing a common pool of genes as a result of interbreeding. It can be characterized according to:

1. size,
2. degree of intactness, that is, how much migration into and out of it occurs, and
3. mating structure, whether random or tending toward inbreeding or outbreeding. For example, if there are social groups between which mating is reduced, such as social classes that do not "mix" much, then the gene pool may differ between classes. Inbreeding customs may also cause local deviations in the gene pool. If major restrictions on interbreeding occur between subgroups of the population, then it may effectively *be* several populations in a genetic sense. Alternatively, if the populations within a species have substantial migration between them, the species as a whole should be considered a single genetic population.

---

or, p² of AA, 2pq of Aa, and q² of aa individuals. (This is the well-known expansion of the binomial [p + q]².) Another way of developing this genotypic array is shown in Figure 16–1.

Suppose we have a million sperm and a million eggs (representing a large population). Let 200,000 of the sperm contain a and 800,000 contain A, and likewise for the eggs. The frequency of a in this gamete sample is thus $\frac{2}{10}$, or q = 0.2, and the frequency of A is $\frac{8}{10}$, or p = 0.8 (p + q = 0.8 + 0.2 = 1). Mixing the gametes together randomly, (0.8 + 0.2)², generates the following array of genotypes: .64AA + .32Aa + .04aa. Note that the gamete frequency of a (q of 0.2) is necessarily different from the genotypic frequency of aa in the next generation (q² of 0.04). What the Hardy-Weinberg model of nonevolution says is that this genotypic array of 640,000 AA, 320,000 Aa, and 40,000 aa individuals will stay the same in succeeding generations (Fig. 16–2).

Given: A million sperm plus a million eggs with the frequency of allele A in the parent population being 0.8 and of allele a being 0.2.

Product: 640,000 AA, 320,000 Aa, and 40,000 aa offspring in a total population of one million.

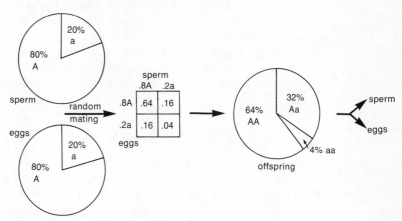

Figure 16–2 Arriving at Hardy-Weinberg genotypic equilibrium in two hypothetical situations.

Given: Sperm and eggs as above with p = .5, q = .5
Product: 250,000 AA, 500,000 Aa, and 250,000 aa offspring

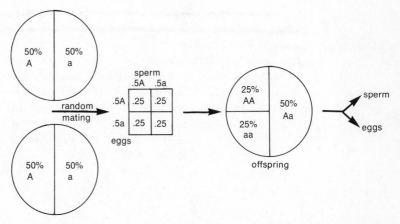

To demonstrate this, let us look at the gametes of these new individuals. To find the total frequency of A-type gametes (p) they will produce, we take all the gametes of the 640,000 AA's and half those of the heterozygotes (of the 320,000 Aa's), or $p_A = p^2 + \frac{1}{2}(2pq)$. This amounts to $\frac{640,000 + 160,000}{1,000,000}$ or $.64 + .16 = .80$. The new p_A is 0.8, just as the previous p_A was. Likewise for the q_a: it is $\frac{1}{2}(2pq) + q^2$ or $\frac{160,000 + 40,000}{1,000,000}$; that is, $.16 + .04 = .20$, the same as the parental value of q_a. Since the gamete, or allele, frequencies of the population are the same as in the previous generation, the offspring produced by these gametes under a regime of random mating will be in the same proportions as before, namely 64% AA, 32% Aa, and 4% aa (Fig. 16–2).

GENETIC DRIFT

Clearly the Hardy-Weinberg model does not describe a real situation for any population of organisms. To begin with, breeding populations are not infinitely large. Particularly in small populations allele frequencies may change rather quickly owing to loss of one allele or another by chance events. Such frequency changes are called genetic drift. Although drift is narrowly defined as a frequency change due to the chance distribution of alleles in meiotic segregation and fertilization (sampling error), drift can be broadly regarded as any accidental change in allele frequency (not due to selection or mutation).

Consider a human population on a Pacific island with 64 AA's, 32 Aa's, and 4 aa's (N = 100 and p = .8, q = .2). Suppose that the 4 aa's were in the same outrigger canoe and were lost at sea in a sudden storm. The population now contains 96 individuals and the allele frequencies on the island are now p = .83 $\left(\text{from } \frac{64 + 16}{96}\right)$, and $q_a = .17$, $\left(\text{from } \frac{16 + 0}{96}\right)$. The change in q is $-.03$. Even this small a change in the genetic makeup of the population, caused in this case by a chance event, has shifted the evolutionary base of the population. The genotype frequencies in the next generation will be different (from [.64 AA + .32 Aa + .04 aa] to [.69 AA + .28 Aa + .03 aa]). The same change would have occurred if the 4 aa's had simply failed to reproduce or if a certain number of Aa's had produced offspring only from their A-containing gametes as a result of the sampling process for alleles during reproduction.

As exemplified in the situation above, chance events tend to lower the frequencies of rare alleles more than they tend to lower the frequencies of common ones. If 4AA's had been lost at sea or if a sufficient number of A-containing eggs had failed to be fertilized, the new p and q would have been .79 and .21, respectively; whereas, when the 4 aa's were lost the new values showed a greater departure from the original .8 and .2, namely p = .83 and q = .17. Chance events, then, can cause the frequency of rare alleles to drift downward with time, eventually causing them to go to extinction (where q = 0) and leaving the population monomorphic for the common allele (p = 1). It is less likely, but still possible, for chance events to cause rare alleles to rise in frequency.

Such, then, is the nature of genetic drift. Many small, isolated human groups have been found to be monomorphic for genes which are robustly polymorphic in large populations. For example, a number of small American Indian tribes are monomorphic for the O blood group. Supposing that their common ancestral population contained A and B alleles, these may plausibly have been lost by the action of genetic drift over many generations. Since about 99 per cent of all human existence has been characterized by small social and breeding groups, some geneticists think that genetic drift has been a major force in the diversification, or microevolution, of humankind. We will consider drift again in the next chapter and compare it with selection at that time.

INBREEDING

Small populations are more at the whim of genetic drift than are large ones, but they are also subject to more inbreeding than are large populations. In a population of 50 or 100 people, inbreeding—the marriage of relatives—is inevitable. In a large population it need not occur; but if the population is class or caste stratified it may be tolerated and in some instances even encouraged (as among some groups in India). The customs of most cultures, though, encourage outbreeding (marriage to nonrelatives). The laws in the United States generally prohibit marriage between persons as closely related as first cousins. In all states a person may not marry a parent, grandparent, child, or grandchild—except in Georgia, where it is specified only that a man may not marry his daughter or grandmother!

In plants and animals, continuous inbreeding produces strains homozygous

TABLE 16–1   Inbreeding*

	AA	Proportions Aa	aa	Allele Frequency (q)
Generation 1: Matings	32 AA × AA	48 Aa × Aa	16 aa × aa	0.4
	32 AA 12 AA		16 aa 12 aa	
Generation 2:	44 AA	24 Aa	28 aa	0.4
	44 AA 6 AA		28 aa 6 aa	
Generation 3:	50 AA	12 Aa	34 aa	0.4
	50 AA 3 AA		34 aa 3 aa	
Generation 4:	53 AA	6 Aa	37 aa	0.4
Generation N:	60 AA	0	40 aa	0.4

*Allele frequencies remain the same over the generations if no evolutionary forces disturb them.

Example: Given a population in the Hardy-Weinberg genotypic proportions of 36 AA + 48 Aa + 16 aa (where p = .6, q = .4), with the population reproducing by self-fertilization the eventual genotypic proportions will be 60% AA and 40% aa, as shown above.

for the maximum number of genes possible, that is, pure strains. The more closely related the parents, the quicker homozygosity is reached. The severest form of inbreeding is self-fertilization, as occurred in Mendel's peas. The proportions of the homozygote types found in an inbred population—whether, say, 10% AA and 90% aa or 50% of each—depend on the allele frequencies. The allele frequencies will remain the same whether a population is inbred or outbred if no evolutionary forces are active (Table 16-1).

NOAH WEBSTER ON INBREEDING*

It iz no crime for brothers and sisters to intermarry, except the fatal consequences to society; for were it generally practised, men would become a race of pigmies. It iz no crime for brothers' and sisters' children to intermarry, and this iz often practised; but such near blood connections often produce imperfect children. The common people hav hence drawn an argument to proov such connections criminal; considering weakness, sickness and deformity in the offspring az judgements upon the parents. Superstition is often awake when reezon is asleep.

For humans, self-fertilization is impossible. Brother-sister matings are the most notorious inbreeding situations among humans. Inbreeding can have harmful effects if a common ancestor of a related couple carried a defective recessive allele which has been passed down to both individuals: then they are both heterozygotes and have a 25% chance of producing a child homozygous for the recessive defect.

---

*Noah Webster, "Explanation of the Reezons Why Marriage Iz Prohibited Between Natural Relations." Collection of Essays and Fugitiv Writings on Moral, Historical, Political and Religous Subjects. Boston: Thomas and Andrews, 1790.

TABLE 16-2   Increased Human Abnormality Due to Inbreeding, Observed in Various Studies*

Country	Type of Defect	Excess of Defects Among Children of First-Cousin Marriages over Controls†
France	Conspicuous abnormalities	3.8
Italy	Severe defects	1.9
Japan	Major defects	1.4
	Minor defects	1.3
Sweden	Major defects	2.3
USA	Abnormalities	2.3

*After Cavalli-Sforza and Bodmer (1971), Table 7.12, p. 368.
†Parents not related.

Since all people, including ancestors, harbor several defective genes as a result of recurrent mutation, inbreeding practices cause real problems, even in large populations (Table 16–2). Inbreeding (and founder effect, see later) also accounts for a wide range of recessive anomalies that are found in small isolated populations throughout the world (see Box on Tristan de Cunha, p. 192). One example of an isolated breeding group in the U.S. is the Protestant religious isolate called the Amish, studied extensively by Victor McKusick. The Amish have a high frequency of otherwise rare disorders: hemophilia, the autosomal form of muscular dystrophy, a severe type of short-limb dwarfism (the Ellis-van Creveld syndrome), and others. Human geneticists often study small isolated groups because they provide a situation for discovering new recessive diseases made "visible" by inbreeding.

MIGRATION

Genes migrate. If 4 aa homozygotes sail to our Pacific island of 64 AA's, 32 Aa's, and 4 aa's, and they settle and reproduce there, the islanders' gene pool will be changed from $q = .2$ and $p = .8$ to $q = .23$ $\left(\text{that is, } \dfrac{16 + 8}{104}\right)$ and $p = .77$. Of course some other island's gene pool lost some a alleles in this maneuver. Moreover, it also lost eight alleles of various sorts for other genes in the migrants' genotypes. Indeed, migration, in contrast to mutation, produces changes in a population's gene pool for all genes at once. And since, at least among humans, family groups usually migrate together (as did European immigrants to the U.S.), the immigrant family is likely to have many genes in common and to be a nonrandom sample of the population it departed from.

FOUNDER EFFECT AND BOTTLENECKS

Founder effect is a special case of genetic drift initiated by an act of migration followed by genetic isolation. If the Pacific islanders we have been talking about had an internecine fracas and one faction left for an uninhabited atoll nearby, it is quite possible that that faction would consist of several families all of which had only AA genotypes. These founders of the new atoll population would then represent a nonrandom sample of the "old country." They would all be AA owing to a founder effect. The founder effect concept perhaps gives us a better picture of the origin of O blood group monomorphism among American Indians than does the notion of genetic drift alone.

If a large population is contracted in size (because of an epidemic, famine, or other considerable disaster), drift will have an opportunity to act powerfully on the small remnant, altering the allele array in the population drastically. This is the bottleneck phenomenon. Although anything that changes gene frequencies is an evolutionary force, bottlenecks and founder effects emerge as forces that accelerate evolution.

TRISTAN DA CUNHA: A HUMAN MICROCOSM

A British study has traced the precarious history of a human population on one of the world's most isolated islands, Tristan da Cunha. It lies about midway between South Africa and Brazil in the South Atlantic. The island environment is harsh. Its volcanic cone has produced a rugged terrain and tall cliff walls around the island. There is little land suitable for farming. Native forms of plants and animals are meager.

Sailing vessels called at the island for water and other provisions for many years before the first settlers arrived. In 1816 British soldiers were garrisoned on Tristan to discourage the French from trying to rescue Napoleon from St. Helena (2400 km. away). One of the soldiers and his family stayed on when the garrison left, and gradually other settlers joined them. By 1827 the little colony was supporting itself by farming and fishing. They traded produce and meat to visiting sailing vessels for tools, sugar, and other items. But as the population grew to 100, the food supply became strained and some islanders emigrated, especially in 1857 and 1858. This reduced the population to about 40 people (a population bottleneck, see below), but it grew again to 100 by 1885, at which time a boatload of 15 islanders was lost at sea while trying to reach a passing ship for the purpose of trade. Because of this tragedy the population increased only slowly from 1885 to 1920. Then it grew rapidly to 267 by the year 1961, despite the occasional exodus of discouraged individuals (Fig. 16–3). In 1961 a larger disaster struck: the volcano erupted and the whole population had to be evacuated to England. The people were returned to the island in 1963.

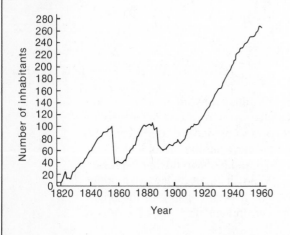

Figure 16–3 Growth of the population of Tristan da Cunha.

During the 145 years before 1961 that this population existed on Tristan, a period of only five generations, it underwent a decline in health and experienced a number of genetic changes. Studies on the Tristan population evacuated to England showed the islanders to be heavily infested with intestinal worms, to be very susceptible to colds and other virus diseases (presumably because of little exposure and thus little immunity to many viruses), and to have a high frequency of birth defects and other genetic abnormalities. The genetic abnormalities included retinitis pigmentosa (a recessive eye defect) and mental retardation.

From available records it was possible for D. F. Roberts, a British geneticist, to study the pedigree pattern of the entire population of Tristan. Such an opportunity arises rarely in human genetics. Roberts showed that only ten of the original 18 adult settlers who had children had descendants living by the year 1961, and that of these ten, two individuals had contributed about twice as much to the 1961 gene pool as had the other eight. These differences in evolutionary success were found to be about 50 per cent due to differences in fertility, and 50 per cent to emigration or accidental death of descendants.

The small size of the population over its 145-year history before the volcanic eruption has meant a dramatic increase in genetic relationship among individuals. Recently, any marriage between islanders has in fact been a marriage between demonstrable relatives. This inbreeding has produced the high frequency of retinitis pigmentosa and, it is suspected, the high frequency of mental retardation (a polygenic condition in this instance). By classifying the 1961 population into six broad categories of mental ability, ranging from normal to very retarded, and comparing the classification with the known degree of inbreeding of the individuals, it appears that those most inbred are generally most retarded and, conversely, those least inbred are the most able intellectually. In addition, the more inbred an individual is, the poorer his vitality and chance of a normal life-span.

This small chapter in human history shows how the genetic foundations of a population are affected by geographic isolation, chance mishaps, and limited resources at odds with population growth.

MUTATION AND SELECTION

New mutations are the basis for evolutionary changes in populations, as mentioned in Chapter Thirteen. In addition, recurrent mutation (from, say, allele A to allele a) can by itself change the frequency of alleles in a population and thus cause a departure from Hardy-Weinberg equilibrium conditions. But mutation, in the real world, is only one ingredient causing changes in gene pools. It is often tempered by adverse selection against individuals carrying the mutation (since most mutations are harmful). On the other hand, the effect of a beneficial mutation can be spread throughout the gene pool by positive selection. That is, if a mutation is beneficial to its bearers they will survive and leave more offspring than individuals lacking it. Selection is discussed in detail in the next chapter.

SUMMARY

A number of forces act on populations to bring about changes in their genetic structure. These evolutionary forces cause departures from Hardy-Weinberg genotypic equilibrium. Among these forces are genetic drift, migration, founder effect, mutation, and selection. Small populations are less buffered against evolutionary change than are large populations.

STUDY QUESTIONS

1. Assuming Hardy-Weinberg equilibrium in a population with $p = .4$ and $q = .6$ in a given generation, what is the expected frequency of the Aa genotype in the next generation?
   (a) .16   (b) .48   (c) .60   (d) .24   (e) none of these

2. Hardy-Weinberg equilibrium in a population is maintained by
   (a) genetic drift
   (b) recurrent mutation
   (c) migration
   (d) all of these
   (e) none of these

3. Evolution requires _____ for selection and other forces to act upon.
   (a) genetic uniformity
   (b) mutants
   (c) environmental uniformity
   (d) genetic diversity
   (e) (b) and (d)

4. Environmental differences in different areas permit _____ to occur.
   (a) recombination
   (b) transformation
   (c) transcription
   (d) fertilization
   (e) local adaptations

5. Evolutionary adaptations of organisms to their habitat
   (a) require genetic diversity
   (b) involve changes in q
   (c) are due to genetic drift
   (d) (a) and (b)
   (e) (b) and (c)

6. Which of the following statements about mutations is false?
   (a) They are the source of inherited differences which can produce evolutionary change.
   (b) They usually result in harm to the organism.
   (c) They are sudden, stable changes in ribosomal RNA.
   (d) They include base substitutions, additions, and deletions.
   (e) They are rare changes in DNA.

7. Hunting-gathering bands of early humans probably
   (a) contained no genetic polymorphisms
   (b) were the victims of genetic drift
   (c) were the victims of continental drift
   (d) contained no polygenic traits
   (e) had no inbreeding

8. "Populations can evolve but individuals cannot." This statement means
   (a) individuals are unable to adapt to changing environments
   (b) populations can change in allele frequencies but individuals cannot
   (c) population size is independent of the genetic code

(d) the number of populations (and species) on earth is constant through time

(e) the number of populations (and species) on earth is constantly changing

9. Does inbreeding by itself change gene (allele) frequencies? Genotype frequencies?

10. Based on the data in Table 16–2, abnormalities arise in the offspring of first-cousin marriages how many times more often than in controls?

   (a) 3.8 times
   (b) 1.3 times
   (c) 1.2 to 3.8 times
   (d) 130 to 380 times
   (e) 13 to 38 times

11. Key terms:

evolution	random mating	gene pool	bottleneck
allele frequency	genetic drift	migration	species
Hardy-Weinberg model	inbreeding	founder effect	

SUGGESTED READINGS

Boyd, W. C. Four achievements of the genetic method in physical anthropology. American Anthropologist 65:243–252, 1963.

Cavalli-Sforza, L. "Genetic drift" in an Italian population. Scientific American 221:30–37, 1969.

Cavalli-Sforza, L., and W. F. Bodmer. The Genetics of Human Populations. San Francisco: W. H. Freeman & Co., 1971.

   An advanced text.

Davis, K. The migrations of human populations. Scientific American 231:93–105, 1974.

Farrow, M. G., and R. C. Juberg. Genetics and laws prohibiting marriage in the United States. Journal of the American Medical Association 209:534–538, 1969.

Harrison, G. A., and A. J. Boyce. Migration, exchange, and the genetic structure of populations. In The Structure of Human Populations, G. A. Harrison and A. J. Boyce (eds.). New York: Oxford University Press, 1972, pp. 128–145.

McKusick, V. A. Genetic studies in American inbred populations with particular reference to the Old Order Amish. In Genetic Polymorphisms and Disease in Man, B. Ramot (ed.). New York: Academic Press, Inc., 1974, pp. 150–158.

McKusick, V. A. Human Genetics. Ed. 2. Englewood Cliffs, N.J.: Prentice-Hall, Inc., 1969.

Pollister, W. S. The physical anthropology and genetics of marginal people of the Southeastern United States. American Anthropologist 74:719–734, 1972.

   Discusses isolates from Florida to Pennsylvania.

Reed, T. E. Caucasian genes in American Negroes. Science 165:762–768, 1969.

   See also Letters, Science 166:1353, 1969.

Roberts, D. F. Genetic effects of population size reduction. Nature 220:1084–1085, 1968.

Roberts, D. F. Consanguineous marriages and calculations of the genetic load. Annals of Human Genetics 32:407–410, 1969.

Sheba, C. Gene frequencies in Jews. Lancet 1:1230–1231, 1970.

Mosquito spirit attacking a human body. Malayan aborigine carving.

A theory is a species of thinking, and its right to exist is coextensive with its power of resisting extinction by its rivals.

T. H. Huxley, Science and Culture (1888)

Chapter Seventeen

DARWINIAN EVOLUTION

THE LIFE AND WORK OF DARWIN

Charles Darwin shocked the world with the idea that life changes, or evolves, over the ages—that new species arise from pre-existing ones as a result of changes in the environment selecting for the survival of new genetic types. This theory was published in 1859 in Darwin's most famous book, *On the Origin of Species by Means of Natural Selection*.

Darwin lived from 1809 to 1882. His father was a well-to-do English physician. Under Queen Victoria, who reigned from 1837 to 1901, England enjoyed immense wealth by exploiting the resources of her colonies. Darwin's position in the upper class enabled him to spend the majority of his life in leisurely study of biological problems, and, for five years, to discover the flora and fauna of distant lands (none far from some outpost of the British Empire).

The riches of the empire fostered in England the liberal belief that humanity was progressing steadily toward material comfort, liberty, and enlightenment (now considered to be a somewhat romantic nineteenth-century notion). Darwin's theory of evolution, which was equated with social progress in the minds of some people, fit very comfortably into this idealistic framework.

Darwin's schooling was of the traditional type, composed largely of the classical languages. He was sent to Edinburgh to study medicine, but after two years he became bored and switched to Cambridge. He stayed there three years, receiving a bachelor's degree in theology although his main enthusiasms were sports, collecting insects and plants, and studying geology. He was fascinated by geology because the Scotsman Lyell, whom he knew, was just then demonstrating the similarities of fossil animals in successive strata in the earth. (It had then been recognized for only several decades that fossils are relics of ancient forms of life.)

At the age of 22 Darwin was offered and accepted a volunteer job as a ship's naturalist on the H.M.S. *Beagle*. While the *Beagle* cruised around the world for five years, principally for the purpose of map making, Darwin observed and collected plants, animals, and rocks from exotic places. Despite his continual seasickness, Darwin made voluminous notes on this trip that were to form the heart of his many books explaining animal and plant diversity and their changes through time.

Previous to his voyage Darwin had passively accepted the traditional dogma of the origin of living species by divine creation and the idea of the constancy of species through time. But he observed during his travels that some kinds of plants and animals were restricted to small areas, and that in different regions one usually found somewhat different species in similar habitats, rather than the same species everywhere. For instance, Darwin spent a month on the Galapagos Islands far off the coast of Ecuador. Very few land animals are established on these volcanic wastelands: five species of reptiles, a few birds, and two species of mammals. Some of these animals are indistinguishable from their counterparts on the South American mainland; others are dramatically different, such as the giant land tortoises found on the islands.

Among the birds Darwin observed in the Galapagos were many species of finches (Fig. 17–1). They differ from each other in appearance, but more importantly they differ in the food they eat (two eat cactus instead of seeds, fruit, or insects) and in their habitats (trees, mangrove shore, etc.). Some of the small islands in the Galapagos group have only one or two kinds of these finches, but all the finches are clearly unlike any birds on the South American mainland (Lack, 1953). Darwin came to think it unreasonable that a divinity had created this separate group of birds for these small, unimportant islands.

The rationale Darwin devised to explain these and other observations was the survival of different hereditary variants in different areas. This idea is known as the theory of natural selection. Darwin thought of such geographically distinct variants as diversifying into separate species over eons of time since no species was thought to have changed into another during the short period of human memory, and geological finds suggested that the earth was very old. (We now know it is over three billion years old, rather than the six thousand years computed from biblical legends.)

Upon returning to England, Darwin married his cousin and settled down to years of analyzing his collected biological specimens and synthesizing his explanations of their distribution around the globe. Ill health, persisting from his seasick days, combined with the family wealth to allow Darwin to lead a secluded life in the countryside as a gentleman scholar. Although he never took up teaching or anything recognizable as a "job," his time was occupied in developing his theory of natural selection and in searching diligently for evidence of its application to all life forms.

He tested the idea of selection by breeding domestic animals. For example, he bred pigeons for many years, constantly selecting (by human design rather than by natural stress) modifications in different lines until distinctive breeds had been established from the original motley (and variable) stock. He reasoned that a similar selective force operated in nature but that the natural mechanism for selection, especially when the population was dense, was competition for food and other resources: the best competitors were the "winners" of maximal survival and reproduction.

Darwin may have gotten the idea of life forms competing for survival from reading an essay on human population problems written by the economist Thomas Malthus. Malthus suggested that mankind multiplies faster than does his food sources. Malthus wrote, "The power of population is indefinitely greater than the power in the earth to provide subsistence to man."

Being a cautious scientist, Darwin published his ideas on natural selection only after a young naturalist, Alfred Wallace, had arrived at similar conclusions independently and was on the verge of publishing them. A mutual acquaintance arranged that the two reports be published together (in 1858). (Recall that competition with others also spurred the "getting into print" in 1953 of Watson and Crick's analysis of DNA, as outlined in Chap. Seven.) In 1859 the publication of Darwin's book *The Origin of Species* elicited both expansive praise (from the liberals) and

Figure 17–1   The 13 finches found on the Galapagos Islands. *1–6,* ground finches; *7–12,* tree finches; *13,* warbler finch. (From Keeton, W. T.: Elements of Biological Science. New York, W. W. Norton and Company, 1967.)

violent condemnation (from the conservatives, who were interested more in moral than in natural principles). The condemnations bothered Darwin only insofar as they dealt rationally with aspects of his theory that species are not unchangeable but do, in fact, evolve. Now most biologists accept the fact of evolution because Darwin's theory of natural selection was meticulously developed and credible. (Likewise Watson and Crick were acclaimed because of their credible and carefully developed theory.)

That Darwin was an outstanding scientist is evident not only because he published the *Origin,* but also because he developed many themes of that book into full-fledged scientific reports, ranging from *On Volcanic Islands* and *On Coral Reefs* (geological works) to *Animals and Plants under Domestication, The Descent of Man, Selection in Relation to Sex,* and *Expression of the Emotions in Man and Animals.* Moreover, Darwin predicted that his theories would throw light "on the origin of man and his history," a prediction that was spectacularly correct since among his followers were the paleontologists who have unearthed plentiful fossils of early man. We now know verifiably that man is descended from an ancient ape.

Indeed, Darwin's evolutionary ideas rendered many observations on nature intelligible. Such disciplines as embryology, anatomy, taxonomy, parasitology, and biochemistry have made better sense when viewed in relation to evolution. Darwin wrote in the *Origin,* "There is grandeur in this view of life, with its several powers, having been originally breathed into a few forms or into one; and that, while this planet has gone cycling on according to the fixed laws of gravity, from so simple a beginning endless forms most beautiful and most wonderful have been, and are being, evolved."

It has been argued that no other scientific pronouncement has had so great an effect on human cultures as did Darwin's pronouncement on evolution—that it redefined human nature and potentialities. His work not only legitimized a broad range of studies on nature, it turned Western culture to a prolonged gaze at the future (away from its musings over the biblical past and ancient Greece). It also deposed man from his medieval-ancient lordliness over nature, placing him, rather, within nature as a biological being resulting from natural processes. Darwin's work showed that the world was not created for humans, but evolved for and of itself. This view made non-naturalistic explanations of man's position on earth unnecessary. Life is seen to proceed as if supernatural explanations never existed, so that science sees no need to adopt them, whether they are true or not. As George Gaylord Simpson (1960) has written, "Evolution is a fully natural process, inherent in the physical properties of the universe, by which life arose in the first place and by which all living things, past or present, have since developed, divergently and progressively" (p. 969).

That Darwinism was later misapplied to justify colonialism, racism, and even genocide—on the argument that the strongest are "fitter" for survival than the downtrodden (Social Darwinism)—is no fault of Darwin's biological theorizing, but is the result of uncritical thinking somewhere in the social sciences.

DARWIN, GENETICS, AND LAMARCK

The reason why certain individuals in nature survive better than others, and as a result produce more offspring, was attributed by Darwin to their superior heredity. However, Darwin did not satisfactorily explain what he meant by heredity. "The laws governing inheritance," he wrote, "are for the most part unknown." Although Darwin and Mendel lived at the same time, Darwin did not know of Mendel's experiments showing that there are discrete genes for different characteristics that obey fixed rules of inheritance.

Of course, Darwin did know, as people have known from time immemorial, that "like begets like"—or at least something similar. Obviously cats beget cats, and frogs, frogs. But people don't beget carbon copies of themselves, though proud parents may insist that such is almost true ("he's just like his father"). Darwin was perplexed by this tendency to variability found in every human family and in many kinds of plants and animals as well.

This was a "hole" in his grand theory: he realized that if all human beings (or all frogs or cats) were genetically alike, there would be no way for them to evolve over long periods of time into something clearly different that could be recognized as a new species—a new kind of animal or plant. Therefore, Darwin argued, it was necessary (as well as true) that there be genetic variability among individuals so that favorably endowed ones could be selectively successful in the environment they all shared. The favorable, or best-adapted, variants would then give rise to more descendants than the others. Such a process would gradually shift the genetic make-up of a population over the ages. One might think this evolutionary process would eventually produce a uniform population, maximally adapted to a certain environment; but this rarely or never occurs, as Darwin and others have pointed out. The reason is that environments are always changing—ice ages come and go, smog descends, a succession of wet years occurs—and thus the hereditary traits suited to survival in the environment shift in turn. Moreover, a desert environment or sea-shore is not uniform throughout its extent. Thus there is spatial differentiation of individuals or groups adapted to particular local conditions. Clearly, then, evolutionary theory must postulate the persistence of genetic variability through time and throughout space in order to explain how groups survive. But since Darwin didn't know about Mendelian genes, how did he explain the persistence of hereditary variants?

About 50 years before Darwin, the French naturalist Lamarck had generalized some traditional beliefs into a theory of the "inheritance of acquired characteristics" that purported to explain adaptive differences between varieties or species. Lamarck thought individual, inherited differences were acquired by life experiences such as the use or disuse of body parts. This idea goes back to Hippocrates and Aristotle in ancient Greece. The fact that a blacksmith's son may have enormous muscles was attributed to his father's having passed on his muscular development via motes of hereditary material that were drawn from his muscular arms into his reproductive cells. (Whether blacksmiths also had muscular daughters is unrecorded by history.) The theory of acquired inheritance is still a popular explanation of heredity in some folk cultures, and until the 1960's was the governmentally accepted theory in the Soviet Union for improving farm crops (Caspari and Marshak, 1965; see Box on Lysenki, p. 202).

But Lamarck's "good sense" theory was not upheld by careful experiments. Before 1900 Weismann showed that cutting off the tails of mice over 22 generations did not produce offspring or descendants that were short-tailed or tailless. In addition, Lamarck's theory of the inheritance of acquired traits was not supported by certain "experiments in nature." For example, deformed or blind persons do not always have similarly defective children. Yet, according to Lamarck, they must needs do so.

LYSENKO

The last holdout against the logic of Mendelian genetics in influential circles was Lysenko, an incompetent agricultural scientist turned bureaucrat in the Soviet Union. Other Russians who were contemporary with Lysenko understood the Mendelian logic and made important contributions to the mainstream of biology. These include N. I. Vavilov (Chap. Eighteen), S. S. Chetverikov, N. Timofeev-Resovsky, and N. P. Dubinin, to name but a few.

Lysenko managed to outlaw genetics in the U.S.S.R. from 1948 until 1964. His own inept theories became accepted dogma. He succeeded in convincing scientifically naive officials, including Stalin, to endorse his theories on the basis of his devotion to communism and his promise to improve agriculture rapidly. As I. M. Lerner (1972) has described the situation, "Recognition of scientific achievement in the Soviet Union, at least through the Stalin era, was less dependent on concrete discoveries and their exploitation than on expression of loyalty to official philosophy" (p. 313).

The chronology of relevant events in Russia is as follows:

1900–
1936   Mendelian genetics versus Lamarckianism is debated among Russian biologists.

1917   Russian Revolution, followed by civil war.

1927   Stalin wins his power struggle with Trotsky, a struggle which had followed Lenin's death in 1924.

1928   Lysenko, then a young man, formulates a nongenetic "theory" of plant development that he maintains will revolutionize agriculture: allowing potatoes and wheat, for example, to grow virtually year round by "training" them to be physiologically adapted to heat and cold (Lamarckianism).

1932   Lysenko starts a scientific journal on plant improvement in which he vilifies scientists who do not praise his work. The propagandist I. Prezent collaborates in these attacks.

1936   At a meeting of the Lenin All-Union Academy of Agricultural Sciences, Lysenko and his supporters attack Vavilov's program of genetic breeding studies to improve agriculture. Vavilov unsuccessfully defends his apolitical scientific methodology. Thereafter Lysenko is appointed to direct Soviet agricultural research, which Vavilov formerly directed. Lysenko promises quick improvements in agricultural productivity.

1938   Stalinist political purges catch many orthodox scientists in their sweep, including Vavilov, who died in Siberia.

1948   Lysenko is entrusted by Stalin with full authority for all Soviet agriculture and most of biology. Criticism of Lysenkoism is forbidden. Biologists and their textbooks are purged wholesale. Scientific discourse with nonsocialist countries is stopped. Mendelian genetics is called reactionary.

1953   Stalin dies. Khrushchev remains supportive of Lysenko.

1964   Khrushchev is deposed because his economic programs are a failure. Soviet agriculture is in obvious shambles. Lysenko is dismissed. Investigations show Lysenko's claims for world-shaking scientific advances to be inaccurate and biased; his experimental results cannot be repeated. Textbooks are revised to include Mendelian genetics and surviving geneticists are rehabilitated.

Lessons from Lysenkoism: (1) Russia was backwards in agricultural research before the 1917 revolution and, thanks largely to the muddling of Lysenko, the U.S.S.R. remained behind in this area into the 1960's. Catching up will take a while yet. (2) Lysenko's abandonment of scientific methodology and verifiability for an extreme "mission-oriented" approach was inevitably self-defeating. The lesson here is that much of "practical" science necessarily rests on the foundations of basic science. (3) The public funding of scientific research, in the U.S. as well as the U.S.S.R., is a constant danger in terms of the possible misuse of scientific knowledge or pseudoknowledge for political ends, whether for personal power or state power.

Darwin accepted Lamarck's ideas to some extent—a fact which tended to publicize Lamarckianism as Darwin's books became "conversational" and controversial. But while Darwin was particularly suspicious of the notion that accidental mutilations were inherited, in some dramatic cases he was almost certain that use "strengthened and enlarged certain parts, and disuse diminished them; and that such modifications are inherited." He contrasted the great development of udders in milk cows and goats with their poor development in non-milked cows as an example of the inheritance of acquired characteristics.

However, on balance, Darwin was inclined to "lay less weight on the direct action of the surrounding conditions, than on a tendency to vary due to causes of which we are quite ignorant." Mendel, of course, was *not* ignorant at that time of the mechanisms of inheritance that could explain this "tendency to vary." The only mechanism of inheritance Darwin could muster up was the erroneous one of blending inheritance, according to which no genetic entity could be imagined to dominate or segregate.

SELECTION ON TRANSIENT AND BALANCED POLYMORPHISMS

Darwin's description of natural selection has been refined in recent decades to mean progressive change through time in the frequency of a genotype. The cause of these changes in frequency is taken to be environmental pressure. If the allele frequency changes from near zero up to 100 per cent in a population, then for each successive generation during the period of such change the *mutant* form must have greater biological fitness: it must be selectively favored by the environment for greater survival and/or greater reproduction than the nonmutant form. In retrospect, we can see how this process of selection might easily have occurred to produce a whole population of wingless flies on wind-swept, isolated islands, and to produce dark-skinned (and thus sunburn-resistant) humans in tropical regions. We cannot so easily guess a Darwinian (selectionist) explanation for the red hair of orangutans—

TABLE 17–1   Natural Selection: How It Works

Given AA, Aa, and aa Genotypes	Selection	End Result
(1) With environment constantly selecting against aa	Unidirectional	$p_A = 1.0$, $q_a = 0$, except for recurrent mutation of $A \rightarrow a$
(2) With environment constantly selecting against Aa and aa	Unidirectional, additive (faster than [1])	$p_A = 1.0$, $q_a = 0$, except as above
(3) Constant selection against Aa and AA	Unidirectional, additive	$q_a = 1.0$, $p_A = 0$, except as above
(4) Constant selection against both AA and aa	Stabilizing (for both alleles)	A and a at intermediate frequencies

but we can imagine that a plausible explanation would present itself if we knew all the ecological and physiological variables.

The type of environmental change we have envisioned so far is a unidirectional one. If we were to catch such a process in action, we would be observing a transient polymorphism, the end result of which is genetic uniformity (p or q being zero for the unfavorable allele; see Table 17–1). If, however, selection occurs simultaneously against both homozygotes, meaning that the heterozygote is the fittest phenotype, a balanced polymorphism is produced with both alleles at intermediate frequencies. The population with this balanced polymorphism would be in equilibrium between two disadvantages. The favored heterozygote is often called heterotic, or overdominant.

Frequency-dependent selection is another possible cause of the maintenance of a polymorphism in a balanced state. In this case, if a trait is rare, it is favored (for example, red hair?), but if it is common, it is not favored. A third means of maintaining such a balanced polymorphism is if the environment is spatially or temporally heterogeneous. Then selection is diversified in direction, depending on which kind of individuals are *where, when.*

In human populations some blood group systems (for example, ABO, MN), some types of hemoglobin (for example, S and C variants), and some types of enzyme defects (for example, G6PD deficiency) are thought to be balanced polymorphisms maintained by opposed selection pressures of some sort. But at least in a few of these cases the polymorphisms may not be balanced; the mutant type may be gradually declining over the centuries as the result of adverse selection now (although selection may have been positive in some previous era). Indeed, it has been shown that among U.S. blacks the frequency of the sickle-cell gene (hemoglobin S) is declining because of the absence of malaria in the United States. In Africa, however, malaria continues to keep pressure on the hemoglobin system, with the heterozygotes—carrying both hemoglobin A and S—being resistant to the disease. There the combination of this heterozygote advantage with the disadvantage of the two homozygous forms (AA being susceptible to malaria and SS being subject to severe anemia) maintains the polymorphism in a balance of allele frequencies (Table 17–2).

DRIFTERS AND SELECTIONISTS

A controversy exists between geneticists who aver that most of evolution is due to selection promoting favorable alleles, and geneticists who maintain that most

TABLE 17–2   Hemoglobin, Malaria, and Sickle-Cell Anemia

Phenotype	Genotype		
	AA	AS	SS
Appearance of red blood cells	Normal	Normal, but when deprived of oxygen they sickle	Sickle-shaped
Appearance of protein band on gel electrophoresis	A ▬	A ▬ / S ▬	S ▬
Composition of protein	Normal	Some molecules are A , some are S	All molecules are S; here valine has replaced glutamic acid (found in hemoglobin A) as the 6th amino acid on the beta chain
Protein formula (Each molecule contains two identical alpha [α] and two identical beta [β] chains)	$\alpha_2^A \beta_2^A$	$\alpha_2^A \beta_2^A$ and $\alpha_2^A \beta_2^S$	$\alpha_2^A \beta_2^S$
Selection on Phenotypes	*Normal*	*Sickle-Cell Trait*	*Sickle-Cell Disease*
Tropical Africa	Susceptible to malaria; Sometimes dies (selective disadvantage of 15%; fitness of 85%)	Resistant to malaria; survives (fitness of 100%)	Usually dies in childhood from sickle-cell anemia (selective disadvantage of 80%; fitness of 20%)
United States	No malaria, therefore lives	Lives	May die, but less chance than in Africa because of medical treatment (disadvantage somewhat less than 80%)
Representative Genotype Frequencies			
Tropical Africa: at birth	68%	30%	2%
as adults	65%	35%	rare
U.S. blacks	90%	10%	rare
U.S. whites	almost 100%	rare	none

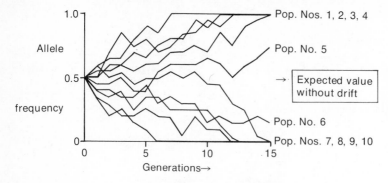

Figure 17–2 Populations of 10 individuals with p = q initially, showing drift of allele frequencies away from 0.5 in different populations at different rates.

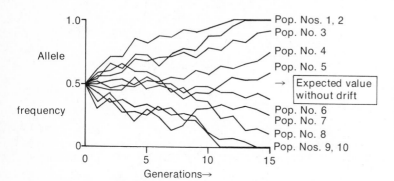

Figure 17–3 Populations of 100 individuals with p = q initially, showing drift of allele frequencies away from 0.5 in different populations at different rates. Compare with Figure 17–2.

of it is due to genetic drift acting on neutral alleles. The term neutral alleles means those that are adaptively equivalent to favorable alleles already existing in a population. For example, a mutant allele that produced a protein as functional in all circumstances as the protein of the "wild-type" allele, even though it had a different amino acid at a certain site in its structure, would be considered neutral in its phenotypic effect. Such neutral alleles are thought by some geneticists to be fairly frequent (as discussed later).

There are two roots to the drift versus selection controversy: one historical and the other resulting from an explosion of new technical information. The historical basis stems from the work of Sewell Wright in the 1930's. Wright pointed out that the random genetic effects of meiosis and fertilization (see Chap. Sixteen) should cause allele frequencies to drift substantially from their initial value in small populations (for example, N = 10) (Fig. 17–2); whereas in medium-sized populations (for example, N = 100) these random fluctuations should cause less drift away from an initial frequency value (Fig. 17–3). Moreover, allele frequencies should drift rather quickly to zero or one in small populations, though rather more slowly to these fixed extremes in larger populations. (But the time of fixation of an allele for either a small or large population is unpredictable in any one case because of the chance events involved.)

Wright's delineation of drift was accepted by his contemporaries in biology but was not considered important in explaining the widespread polymorphisms then known for humans (mainly blood groups, skin color, and a few genetic diseases). Balancing selection was held to be the force maintaining these important variable traits.

By the 1960's, however, many more polymorphisms had been discovered in human and other populations (Table 17–3). Most of this new information on protein diversity comes from the use of the technique of electrophoresis. By this tech-

TABLE 17-3   Genetic Variation in the Human Species: Gene Frequencies of Some Major Polymorphisms

	Allele	Northern Europe	Sub-Saharan Africa	China
ABO blood groups	A	0.28	0.18	0.19
	B	0.06	0.11	0.17
	O	0.66	0.71	0.64
MN blood groups	M	0.54	0.58	0.61
	N	0.46	0.42	0.39
Rh blood groups	R	0.61	0.81	1.00
	r	0.39	0.19	0
Acid phosphatase	P^a	0.36	0.16	0.22
	P^b	0.60	0.81	0.78
	P^c	0.04	0.01	0
6-Phosphogluconate dehydrogenase	A	0.98	0.94	0.93
	C	0.02	0.06	0.07
Phosphoglucomutase$_1$	PGM^1	0.76	0.78	0.75
	PGM^2	0.24	0.22	0.25
Haptoglobin*	Hp^1	0.38	0.56	0.28
	Hp^2	0.62	0.44	0.72

*An iron-binding protein in blood serum.

nique a mutant protein with an electrical charge different from that of the "wild-type" protein will form a separate band in a gel when treated with an electrical current (see Fig. 10–3 and Table 17–2).

The neutralist school asked the question: why are these electrophoretic variants at significant (that is, polymorphic) frequencies in the wide gamut of species studied by this new technique? Although they recognized that many new mutations are harmful and are eliminated or kept at minuscule frequencies in populations by natural selection, researchers like Kimura suggested that the protein polymorphisms revealed by electrophoresis are too numerous to be explained by selection. On the average, an individual is heterozygous at seven to ten per cent of his or her loci (according to different studies), and 30 per cent of all genes studied are polymorphic. If all these polymorphisms are maintained by balancing selection, then this level of polymorphism would impose a significant genetic weakness on many individuals and a "genetic load" on the whole population. Genetic load refers to an average group incapacity or group "cost," while selective disadvantage refers to an incapacity of an individual relative to others in a group. (See Box on Genetic Load, p. 209.)

Because of these considerations on genetic load, the neutralist school says that most protein polymorphisms are due mainly to genetic drift. The selectionists, such as Ayala, maintain that existing polymorphisms are due to selection and regard drift as having a minor evolutionary role, chiefly in special cases of founder effect or bottlenecks in population size. Selectionists further maintain that the argument on genetic load may be invalid. Thoday (1975) has pointed out that truly neutral mutations include only those that involve two kinds of codon changes (Chap. Eight): (1) changes to another synonymous codon for the same amino acid, such as switches

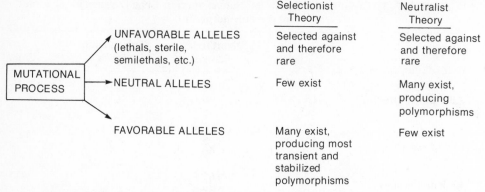

	Selectionist Theory	Neutralist Theory
UNFAVORABLE ALLELES (lethals, sterile, semilethals, etc.)	Selected against and therefore rare	Selected against and therefore rare
NEUTRAL ALLELES	Few exist	Many exist, producing polymorphisms
FAVORABLE ALLELES	Many exist, producing most transient and stabilized polymorphisms	Few exist

Figure 17–4 Summary of drift versus selection theory.

among the six codons for serine (although Richmond [1970] has shown that not even all synonymous mutations need be neutral); and (2) changes in codons such that one amino acid in a protein is substituted for another of similar shape and charge so that there is no effect on the functioning of the protein involved. However, a mutation of one allele to another may have a neutral effect on fitness under present conditions, but an adverse (or favorable) effect under future conditions. Such conditionally neutral mutant alleles may abound in nature as a result of being favored in the past, of being favored only at one stage of development, or of being favored only in certain seasons of the year. Conditionally neutral polymorphisms would seem to be a component of long-term fitness for a species, providing adaptation to environmental changes.

Although the big question between the neutralists and selectionists is what maintains the many polymorphisms now known (Fig. 17–4), from the viewpoint of evolution it is the adaptation of a population to its environment (and thus selection) that is important for survival. Among all the evolutionary forces we know about, selection alone can produce genetically adaptive change. Genetic drift is just "noise" in this system. As one geneticist put it: if most polymorphisms are due to drift, we are interested in the rare ones that are due to selection, and if most are due to selection, we are still interested in those due to selection.

SUMMARY

In Europe through the Middle Ages all creatures were thought to have been put on earth by acts of God and to be unchangeable. Lamarck suggested that these creatures, however they were put on earth, change through time by inheriting traits which arise as a response to the animal's needs in its environment and are likewise preserved by the process of use versus disuse of body parts.

Darwin also thought that creatures change through time. He agreed with Lamarck that variant traits may arise by use versus disuse, but he maintained that such traits were preserved in a population by a different process: natural selection.

Although Mendel did not ask how variant traits arise in the first place, he demonstrated how organisms can vary phenotypically as a result of genetic recombination mechanisms.

We now recognize the interplay of drift, various kinds of selection, and other components of the evolutionary game, but we also acknowledge that many questions about evolution are still unstudied and unanswered.

GENETIC LOAD

Genetic load is defined as a loss in the average fitness of a population due to selection against (1) deleterious homozygotes resulting from inbreeding, or (2) deleterious homozygotes segregating out from a balanced polymorphism, in either an inbreeding or a randomly mating population.

Deleterious alleles arise by mutation so, particularly in (1) above, selection against homozygotes depends not only on the extent of inbreeding, but also on the mutation rate (mutation-selection balance). Environmental mutagens, a concern nowadays, can raise mutation rates and thus increase the genetic-mutational load. Other components of genetic load have also been identified.

STUDY QUESTIONS

1. Charles Darwin is famous for his contribution of the idea of _____ to human understanding of nature.
   (a) changes in gene frequency
   (b) evolution by genetic drift
   (c) biological unchangeableness
   (d) adaptive change in populations through environmental stress on genetic variants
   (e) allele segregation
2. Many genetic variants or defects of red blood cells have been discovered. One reason for this is that
   (a) without hemoglobin, people would die
   (b) red cells have more diseases than other cells do
   (c) all of these variants produce immunity to malaria
   (d) blood transfusions have become common
   (e) it is easier to get blood samples from people than samples of other tissues
3. A biologist thinks of a population as a
   (a) Watson-Crick group
   (b) breeding group
   (c) homozygous group
   (d) Hardy-Weinberg group
   (e) political group
4. Proteins of human red blood cells
   (a) are of two possible kinds: hemoglobin S vs. hemoglobin A
   (b) may be important in natural selection in humans
   (c) may be synonymous codons
   (d) bear no relationship to genetic drift
   (e) are among the most difficult human traits to study genetically

THE POKER CHIP GAME OF POPULATION GENETICS

To demonstrate that genetic drift allows fluctuations in the proportions of genotypes from generation to generation in finite populations even if mating is supposedly random, entertain yourself and ten or more fellow students with the following exercise:

Step A. Each player takes two poker chips at random from a pile of chips of two colors (say, 2/3 red [R] and 1/3 white [W]).

Step B. Tally the genotypes (RR, RW, or WW) of each player for "generation 1" on a score sheet set up as shown in the example below, and calculate the (allele) frequency of red chips in the gene pool.

Players

	1	2	3	4	5	6	7	(etc.)	Frequency of red chips
Generations 1									
2									
3									

Step C. At the call of "Mate!" from the designated genetic leader, all players (parents) mate once by holding out their hands, with one chip hidden in each, toward a partner chosen at random in the group.* Each partner of the "mating pair" taps one hand of his mate and receives the chip in that hand. Parents are now magically transformed into offspring. (If there is an odd number of players, then one person is "unmated" in that generation; but such an individual is available for "mating" in subsequent generations.)

Step D. Tally the new genotype distribution on the line for generation 2 and recalculate the red chip frequency.

Step E. Continue the call of "Mate," the random mating, and the tallying for two or more generations.

Question: Did the red chip frequency stay constant down through the generations? Why, or why not?

Step F. To show the action of adverse selection, have players keep their chips at the end of the last generation tallied above. Now, suppose that the RR combination is susceptible to death by malaria (as is the case for people homozygous for hemoglobin type A). Suppose further that the WW and WR combinations are equally viable and equally resistant to malaria (an assumption contrary to what we know about hemoglobin S and sickle-cell

---

*Each individual here is a hermaphrodite that engages in mutual cross-fertilization, a condition known to occur in garden snails and earthworms, but not in humans.

anemia). Now let the leader show "Malaria!" The people holding RR then fall out of the game. Upon the call of "Mate" the survivors exchange a chip as before. Calculations of the red chip frequencies over several generations should now show the effects of selection against RR.

Question: Do you expect the R allele to be completely eliminated from the population? If so, when?

5. Some of the chemicals that humans put into the environment today (insecticides, food additives, and so on) are mutagenic in humans and other organisms. Thus technology
   (a) is a cause of change in gene frequencies
   (b) may be stopping evolution in our species
   (c) is a factor in the genetic-environmental interaction that relates to the genetic load
   (d) (a) and (c)
   (e) (b) and (c)
6. How does knowledge of DNA and RNA relate to Charles Darwin's evolutionary ideas?
7. According to L. C. Dunn (1965), Darwin crossed normal snapdragon flowers with those that were peloric (Fig. 17–5). He grew the F_1 seed in "two great beds" and none of the plants produced peloric flowers. He let some of the F_1's self-cross and observed in the F_2 37 peloric to 90 normal. Darwin deduced nothing significant from this. Comment.
8. Present a selectionist-Mendelian explanation of the (presumed) muscular prowess of blacksmiths and their sons, and show how it differs from the selectionist-Lamarckian explanation given in the chapter.
9. Darwin did not know of Mendel's work. If he had, do you think it would have made any difference in his theorizing?
10. Chance is important in evolutionary theory in terms of
    (a) genetic drift

Figure 17–5

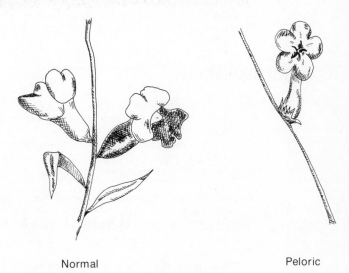

Normal   Peloric

Snapdragons

(b) the timing and direction of mutation

(c) natural selection

(d) the origin of particular combinations of genes

(e) (a), (b), and (c)

(f) (a), (b), and (d)

11. List the factors causing the childhood deaths tabulated* below:

City	Relationship of Parents		
	Unrelated	Second Cousins	First Cousins
Hiroshima	3.6% deaths (1384)†	4.4% (230)	6.1% (532)
Nagasaki	3.4% (2078)	3.2% (330)	5.3% (826)

*From W. Schull and J. Neel: The Effects of Inbreeding in Japanese Children. New York: Harper & Row, 1965.

†Numbers in parentheses indicate number of families studied.

12. Key terms:

evolution

natural selection

Darwinism

adaptation

Lamarck

transient polymorphism

balanced polymorphism

stabilizing selection

biological fitness

heterozygote advantage

neutral allele

REFERENCES*

Caspari, E. W., and R. E. Marshak. The rise and fall of Lysenko. Science 149:275–278, 1965.

Dunn, L. C. A Short History of Genetics. New York: McGraw-Hill Book Co., 1965.

Lack, D. Darwin's finches. Scientific American 188:66–72, 1953.

Lerner, I. M. Dialectical materialism and Soviet science. Quarterly Review of Biology 47: 313–316, 1972.

Richmond, R. Non-Darwinian evolution: a critique. Nature 225:1025–1028, 1970.

Simpson, G. G. The world into which Darwin led us. Science 131:966–974, 1960.

Thoday, J. M. Non-Darwinian "evolution" and biological progress. Nature 255:675–677, 1975.

SUGGESTED READINGS

Appleman, P. (ed.). Darwin. New York: W. W. Norton & Co., Inc., 1970.

An anthology of pre- and post-Darwinian writings, as well as selections from Darwin himself.

Ayala, F. J. Biological evolution: natural selection or random walk? American Scientist 62: 692–701, 1974.

Baer, A. S. (ed.). Heredity and Society. Ed. 2. New York: Macmillan Publishing Co., Inc., 1977.

Bajema, C. J. (ed.). Natural Selection in Human Populations. New York: John Wiley & Sons, Inc., 1971.

Caspari, E. W., and R. E. Marshak. The rise and fall of Lysenko. Science 149:275–278, 1965.

*Note: items cited in the References are also recommended for suggested reading.

Explains the Larmarckian pseudoscience that misled Soviet agriculture for 30 years. See also Lerner, under References.

Darwin, C. Origin of Species. New York: Mentor Books, Imprint of New American Library, 1958.

Originally published in 1859 in London.

deBeer, G. The world of an evolutionist. Science *143*:1311–1317, 1964.

Gish, D. T. Creation, evolution, and the historical evidence. American Biology Teacher *35*: 132–140, 1973.

An interesting anti-evolution review.

Grabiner, J. V ., and P. D. Miller. Effect of the Scopes trial. Science *185*:832–837, 1974.

Discusses the famous Tennessee trial of 1925 on teaching evolutionary theory in the classroom. (See also comments by G. G. Simpson, Science *187*:389, 1975.)

Huxley, T. H. Science and Culture. London: Macmillan, 1888.

Kolata, G. B. Population genetics: reevaluation of genetic variation. Science *184*:452–454, 1974.

Mayr, E. The nature of the Darwinian revolution. Science *176*:981–989, 1972.

Moore, J. A. Creationism in California. Daedalus *103*:173–189, 1974.

Lucidly discusses some recent events in the U.S. relating to Charles Darwin.

Clay model of cow barn found in an Egyptian tomb dating from dynastic times. From Metropolitan Museum of Art, Museum Excavations, 1919–1920; Rogers Fund supplemented by contribution of Edward S. Harkness.

At the forefront of the Green Revolution is the genetic development of new crop varieties with higher yields. That some of these are less nourishing, and most poorer tasting, may be dismissed as the necessary price of more carbohydrates and fats for a growing population. Each new miracle plant does not simply replace a less efficient form; it replaces hundreds and hundreds of varieties. Originally domestication was a local affair. Plants adapted to particular soils and climates were cultivated in each area. Hybrids of the domestic plant and its wild relatives sprang up along field boundaries, a rich reserve of genetic variability. . . .

In 1930, 80% of the wheat grown in Greece was composed of scores of local varieties. By 1966, less than 10% were native strains. The rest was a single imported super-plant. The biology of evolution, the principles of ecology, and the history of agriculture guarantee that sooner or later a disease will devastate that alien. Where previously diseases were local and food shortages in any year equally restricted, super-crops and massive famine are natural partners.

Paul Shepard, The Tender Carnivore and the Sacred Game (1973)

Chapter Eighteen

AGROGENETICS

Much of the early work on genetics was directed toward solving practical problems in agriculture. The rediscoverers of Mendel's work in the year 1900 had strong interests in plant breeding; many of the early geneticists in the U.S. worked at government-sponsored agricultural experiment stations; and some university departments of genetics in the U.S. developed from an agricultural background. Studies on domestic animals and crop plants have not only led to their genetic improvement, but have also contributed to our understanding of basic genetic mechanisms. The "pure" and applied aspects of agrogenetics are historically inseparable.

One early study on wheat provided evidence for complex multigenic inheritance of a trait (see also Chap. Fourteen). Nilsson-Ehle showed in 1912 that the color of wheat seeds could be explained as the result of three independent genes, each with a dominant allele which had a reddening effect on seed color (see Box, p. 216). When he crossed an inbred white-seed variety with certain inbred red strains, the F_2 showed conventional 3:1 ratios for red versus white. But when he crossed the white variety with other inbred red varieties he got F_2 ratios more slanted toward red (15:1 and 63:1), indicating that two additional genes were involved in seed color determination.

The genetics of seed color (kernel color) for corn was found to be even more complex than for wheat. Three genes determine whether there will be any color at all in the kernels, and a fourth gene controls whether the color that *is* produced will be purple or red (as occurs in "Indian" corn).

Corn was extensively studied by plant geneticists in the U.S. because it was a native agricultural staple and existed in many varieties, thanks to countless generations of selective breeding by American Indians. Their systematic experimentation over the centuries produced over 300 varieties of corn—some specialized for growing conditions as far north as the Gulf of St. Lawrence, some drought-resistant varieties specialized for the deserts of the Southwest, and others adapted to other areas.

The U.S. corn belt's fame rests on hybrid corn. Hybrid corn was developed in 1917 as a result of a fortuitous combination of research findings by a small number of plant geneticists. It was vastly superior to traditional varieties in yield and vigor.

215

THREE GENES FOR SEED COLOR IN WHEAT

Alleles a, b, and c are recessive to A, B, and C, respectively. At least one dose of A, B, or C is necessary for red color in wheat seeds. Thus genotype aa bb cc has white seeds, and all other possible genotypes are red (A– B– C–, aa bb C–, etc.).

Given one inbred white-seed variety: (1) aa bb cc

and three inbred red-seed varieties
- (2) AA bb cc
- (3) AA BB cc
- (4) AA BB CC

1. A cross between white (1) and red (2) (aa bb cc × AA bb cc) gives an ordinary 3:1 ratio of red to white in the F_2 (from Aa × Aa).

2. A cross of white (1) and red (3) (aa bb cc × AA BB cc) gives an extraordinary ratio of 15 red to 1 white in the F_2 (from Aa Bb × Aa Bb), where $\frac{15}{16}$ are red because they possess at least one capital allele (A– B–, aa B–, and A– bb), and $\frac{1}{16}$ is white (aa bb).

3. A cross of white (1) × red (4) (aa bb cc × AA BB CC) gives an extraordinary ratio of 63 red to 1 white in the F_2 (from Aa Bb Cc × Aa Bb Cc).

Paul Mangelsdorf notes that about 25 men, the developers of hybrid corn, ". . . played a major part in an agricultural revolution affecting more than 150 million Americans. . . . [They made] an important contribution toward determining the prosperity and destiny of the entire group."

Before World War II, but particularly during it, hybrid corn swept over the corn-growing area of the U.S. Many kinds of hybrid corn were developed to fit growing conditions in different regions. By 1949 hybrid corn accounted for 78 per cent of all corn grown in the U.S. The food-supply advantage of this shift to hybrid corn was real, but disadvantages are now also apparent (see below).

Much of the farm-planted corn is actually a double hybrid: it is fashioned by separately hybridizing two inbred lines (say, line A × line B) and two other lines (say, line C × line D), and then crossing the AB hybrid to the CD one to produce the seed that is planted for the corn crop (Fig. 18–1). The result of this double hybridization process is a plant with a greater yield of corn than either the inbred lines or even the single-cross hybrids can muster up. The hybrid vigor, or heterosis, of the double-cross plant is world renowned. By maximizing heterozygosity for the gene makeup of the plant, the double crossing technique is thought to permit optimal flexibility of enzyme action for the development of luxuriant growth. But the obvious value of this hybrid vigor rests on the expensive and nonobvious first step in corn production—inbreeding for several generations until the plants are

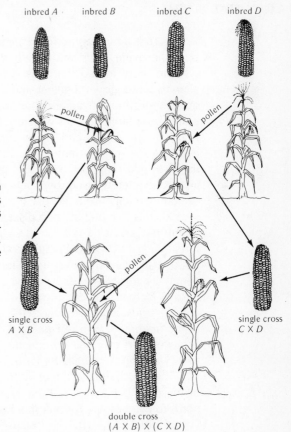

Figure 18–1 The production of hybrid corn from four inbred lines, A, B, C, and D. Paired crosses between the inbred lines produce two vigorous hybrid plants, AB and CD, which are then inter-crossed to yield the double-cross hybrid, ABCD. (From Strickberger, M.: Genetics. New York: The Macmillan Company, 1968.)

spindly, with paltry ears and few seeds. Paradoxically, by "advancing backwards" via these largely homozygous inbred lines, it is then possible to produce a sturdy hybrid of great merit.

Hybrid corn served as a model of success for the subsequent development of a profusion of hybrid crops, from radishes and tomatoes to wheat and zucchini. Hybrid chickens also came into being. Even roses and other ornamental plants are now hybridized by plant growers. All this stands in sharp contrast to the traditional ways of many breeders from time immemorial—inbreeding selectively to produce a "pure" race of, say, Guernsey cattle, Irish setters, or Siamese cats. The trouble with these inbred populations is that they tend to be less vigorous than mongrel ones or deliberately hybridized types. However, hybridization between different races or varieties does also have an ancient aspect: traditional beef cattle, wheat plants, and potato plants, to name but a few, are known to be venerable hybrids.

It should be pointed out at this juncture, however, that the effectiveness of the intensive agriculture of our technological society does not depend only on hybridized genotypes. It depends also on heavy doses of fertilizer, on irrigation (in some areas), pesticides, heavy agricultural equipment (for tilling, sowing, and harvesting), and on an efficient system of storage and transport for farm products. How some of these factors interact with genetics in present-day societies is a topic for later discussion. At the moment we will consider another difficulty with intensive agricultural practices: the vulnerability of commercially desirable crop genotypes to diseases which produce disastrous losses when such plant types are cultivated on a large scale.

CORN LEAF BLIGHT

Corn is a hermaphrodite. The tassels atop the plant bear the pollen and the ears bear the ovules. It is traditionally wind-pollinated. So, in fact, "pre-genetic" corn was not very inbred, but was quite heterozygous because plants from different sources were grown together and wind-carried pollen brought about a mixture of cross-fertilization. Traditionally, then, every plant was distinct from neighboring plants for a large number of genes. Not so today: the plants in a corn field are quite uniform genetically, and, moreover, corn monoculture often extends as far as the eye can see over farming areas of the U.S. And although hybrid corn throughout its cultivated range in the U.S. is not a single heterozygous genotype, because of locally adapted types, most of our corn has been selected for uniformity for a number of commercially useful genes.

All this is a prelude to a discussion of a fungal (mold) disease called corn leaf blight that struck the U.S. corn crop in nearly epidemic proportions in 1970. This threat of destruction to a major food staple was the direct result of genetic uniformity imposed by our intensive agricultural practices. The uniformity at fault involves a genetic factor called Texas male sterility. The male-sterile trait was bred into many lines of corn to prevent unwanted self-fertilization during the production of single-cross hybrids. Rows of male-fertile plants were grown among the male-sterile ones in the field, as shown in Figure 18–1, so that controlled cross-fertilization could be achieved. Prior to the introduction of male-sterile strains, detasseling of certain rows was done manually by farm workers, a laborious and expensive process.

The fungal disease under discussion causes rotting of the stem, stalk, and leaves of the corn plant. In 1970 this blight infection of corn was rampant from Florida north to Minnesota. The infestations destroyed 15 per cent of the U.S. crop (about 700 million bushels*). The disease spread northward from the southeastern states by fungal spores wafted on favorable winds. It spread easily throughout the corn belt because 80 per cent of the farm-planted corn carried the male-sterility factor, and this made them susceptible to the parasitic fungus. Farms in the southern Gulf states commonly lost over 50 per cent of their corn crop. The chemical basis of the plants' susceptibility has not yet been discovered, but it has been found that the fungus produces a toxic substance which causes abnormal functioning of the mitochondria in blight-susceptible, but not blight-resistant, plants.

A question that comes to mind at this point is: why wasn't the hybrid corn developed by plant geneticists tested for vulnerability to this disease before it was released to farmers? The answer is that it *was* tested. In all crop-breeding programs, plant strains are routinely tested for resistance to parasitic infections before their seed is released. But new mutant phenotypes of pathogens do appear (see Box on Host-Parasite Relations, p. 219). This seems to be what happend on the corn blight situation; the blight organism mutated and the mutant form was selectively favored. Even then the disease would not have spread so far and done so much damage if the male-sterile phenotype had not been so extensively adopted throughout the corn belt in 1970. Fortunately, by the time the 1971 crop was to be planted considerable quantities of blight-resistant seed were available from seed producers. Some farmers, however, had to switch to growing sorghum or soybeans in 1971 to avoid the possibility of substantial crop loss.

---

*1 bu = .036 cubic meters.

HOST-PARASITE RELATIONS IN PLANTS

Plant diseases are caused by insects, molds, viruses, and other factors. Almost all we know about plant diseases comes from studies on farm plants because of the constant need for practical applications of such knowledge. The susceptibility or resistance of plants to disease organisms is due to the action of a number of genes. Whether the disease organisms are virulent or non-virulent for particular plants is also due to the action of certain genes. The genetic situation then must take into account the interaction between the plant and the parasite, as indeed is true of any host-parasite relationship (even for humans). Precisely which genes of the parasite react in what way to specific plant host genes is not yet known for any crop plant, despite prolonged research.

Breeding for plant resistance to diseases, then, is done empirically, on the basis of what "works" rather than on the basis of what should work, given an understanding of the genetic-metabolic relations involved. The trouble with this empirical approach of the plant geneticists is that new genotypes of parasites arise regularly by mutation or genetic recombination, rendering the previously resistant plant susceptible to the new forms. Thus, for example, wheat breeders in the U.S. must constantly produce new wheat strains for farmers that are resistant to newly arisen strains of "stem rust" (a parasitic mold). The "lifespan" of disease-resistant wheat varieties in the U.S. is only five years. Most major crops are in a continual battle, or occasionally an uneasy truce, with their plant diseases. As the host changes, the parasite manages to change too; they are thus said to coevolve over long periods of time. The plant geneticist faces a never-ending problem in this regard.

Will it ever happen again? Undoubtedly so. There have been rampaging plant diseases before, and there is no reason to think they will not recur. In 1954 a stem rust disease of wheat in the U.S. destroyed 75 per cent of the crop of hard wheat (used to make spaghetti and other pasta products) and 25 per cent of the bread wheat. Plagues of locusts have occurred repeatedly, at least since biblical times. Such plagues, or any of the currently minor diseases of wheat or corn, could again flare up to major proportions if the right mutations occur, if the weather is favorable, and if dense plantings of susceptible hosts are available for attack. It is well to reflect that humans are in constant competition with numerous other life forms for what we possessively think is our own personal food supply.

THE GREEN REVOLUTION

Plants are emphasized more than animals in this chapter because plants are the basic source of all our food. Most people of the world are largely vegetarian (Table 18–1). India and China are able to sustain their large populations because their people eat mainly fruits, vegetables, and grain products. The energy in plant

TABLE 18–1   Major Food Crops of the World*

Grains	Root Crops	Fruit Crops	Legumes	Other
Rice	Cassava	Coconut	Mung bean	Sugar cane
Wheat	Sweet potato	Banana	Soybean	
Corn	Potato		Cowpea	
Sorghum	Sugar beet		Chick and	
Millet			pigeon pea	
Rye			Peanut	
Barley				

*These crops furnish more than 90% of the world's food and account for three-fourths of the world's cultivated acreage (after Wittwer, 1975).

foods is more efficiently utilized by eating the plants directly rather than letting a cow eat them and then eating the cow. The cow "wastes" a lot of the plant material in making hooves, hide, and other parts that are inedible; it also wastes plant material by respiring (breaking down sugars to carbon dioxide and water), moving around, and excreting some of the materials. In general, farm animals must be fed three to ten pounds* of grain to produce one pound of meat; for one pound of beef the figure is seven pounds of grain. According to a report by T. Poleman (1975), an average person in the U.S. now "consumes" 1800 pounds of grain per year, but only 100 pounds of that is consumed directly; most of the remaining 1700 pounds is fed to animals.

The term green revolution came into vogue in the 1960's when high-yielding genotypes of wheat and rice were developed and seemed to be the panacea for the problem of feeding the growing world population, particularly the large part of it found in poor countries (Fig. 18–2). But by 1970 it was clear, even to the leaders of the green revolution, that the increased yield of the new varieties only bought the world some time, so that the "population bomb" would not explode until 1985 or 1990.

The intensive agricultural practices of the U.S. and other Western countries rely on the heavy use of fertilizer, expensive irrigation schemes, farm machinery, and so on. These props were also advocated by plant breeders for optimal productivity of the new high-yielding varieties when they were introduced into the

*1 pound = .45 kilogram.

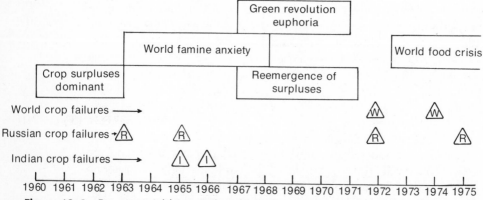

Figure 18–2   Recent social history of world food. (Based on Walters, H.: Difficult issues underlying food problems. Science, *188*:524–530, 1975.)

		Diseases					Insects			Soil problems			
	Lodging	Blast	Bacterial blight	Bacterial leaf streak	Grassy stunt	Tungro	Green leaf-hopper	Brown plant-hopper	Stem borer	Alkali injury	Salt injury	Iron toxicity	Reduction products
IR8	R	MR	S	S	S	S	R	S	MS	S	MR	S	MR
IR5	MR	S	S	MS	S	S	R	S	S	S	MR	S	MS
IR20	MR	MR	R	MR	S	R	R	S	MS	S	MR	R	MR
IR22	R	S	R	MS	S	S	S	S	S	S	S	MR	MR
IR24	R	S	S	MR	S	MR	R	S	S	MR	MR	MR	MS
IR26	MR	MR	R	MR	MR	R	R	R	MR	MR	MR	R	MR

Figure 18–3 Rice varieties released by the International Rice Research Institute each year are continually improved in their genetic endowment. IR–8 was the first variety released, IR–26 the most recent. R, resistant; MR, moderately resistant; MS, moderately susceptible; S, susceptible. (From IRRI Research Highlights for 1973.) Lodging refers to a plant with a weak stem toppling over when it has a heavy head of grain on it. (From Wade, N.: Green revolution. Science, *185*:844–845, 1974.)

farming economies of poor countries. While some critics of the green revolution said that the new varieties only showed high yields if lavished with fertilizer, water, and pesticides, defenders of the "miracle grains" claimed that the new varieties also grew better than traditional ones even without intensive care. One indication of this is that recently developed varieties of rice, for example, are resistant to a wide range of diseases and growing problems (Fig. 18–3). But a little reflection on the corn blight saga will suggest that monoculture of a disease-resistant rice will strongly select for new genetic strains of its parasites, a situation that will require the development of new rice strains.

The anti-green revolution forces became vociferous when it was obvious that the new grain varieties would not adequately feed the world without enormous expense in the "backup" sector. And even to keep future production on a level with that attained in, say, 1970 will be difficult because of the increasing shortages of fuel and fertilizer in the world. (Most nitrogen fertilizer is a petroleum product.) Moreover, the critics of the green revolution pointed out that it did not work well in the "third world" because, in addition to the problems of expense, there were cultural problems in introducing the varieties. Some of them were regarded as unpalatable and their introduction often required practices at odds with religious views of tilling the soil. They benefited rich farmers more than poor ones (who can ill afford seed and fertilizer), and caused social strife by benefiting areas with good land but not neighboring areas with marginal land. And because storage, transport, and other secondary problems of the new agriculture had been naively ignored by the Western-based agricultural planners, charges of agricultural colonialism were made in some countries: fertilizer, well pumps, farm tools, and other needed paraphernalia often had to be bought by poor countries from Western ones at what seemed exorbitant prices.

In addition, the real but modest increases in grain yield occasioned by planting the new varieties of wheat and rice in poor countries had, in turn, led to an increased population for them (Table 18–2). Not only has the death rate among some segments of such populations decreased, but the birth rate has spurted and caught up with the increased food supply. The problem is how to speed up the production of food and simultaneously slow down population growth. This problem is

TABLE 18–2   Total Indian Food Grain Production, Imports, Changes in Government Stocks, and Net Availability*

Year	Production ($\times 10^6$ tons)	Net Imports ($\times 10^6$ tons)	Population ($\times 10^6$)	Net Availability Per Capita (kg/year)
1950–1951	50.8	4.80	363.2	134.0
1955–1956	66.9	1.44	397.3	152.5
1960–1961	82.0	3.50	442.4	170.5
1964–1965	89.4	7.46	482.5	175.4
1965–1966	72.3	10.36	493.2	145.6
1966–1967	74.2	8.67	504.2	146.5
1967–1968	95.1	5.70	515.4	168.6
1968–1969	94.0	3.87	527.0	162.5
1969–1970	99.5	3.63	538.9	166.2
1970–1971	108.4	2.05	550.8	171.3
1971–1972	105.2	−0.48	562.5	172.8
1972–1973	97.0	4.12	576.7	153.1
1973–1974	103.6	6.10	589.7	167.0

*Data from Bulletin on Food Statistics (New Delhi: Government of India, Ministry of Agriculture, 1973, p. 170), and Draft Fifth Five Year Plan (New Delhi: Government of India, Planning Commission, 1973, Vol. 1, pp. 1–2). From J. D. Gavan and J. A. Dixon (1975).

now, finally, being recognized by rich nations and international agencies in their planning of economic assistance to poor nations. But the time is very late. At present growth rates, the world population will double again in another 30 to 40 years, and no one is predicting that the food supply will double by then. Although there is enormous room for improvement in the food supply, especially in the starchy root crops that are the staples of the tropics (Fig. 18–4), future gains in the food supply will be even more expensive than heretofore. The best land is already under intensive cultivation and the energy crisis is expected to be long lasting.

Criticism of the green revolution has now spread from the political problems it has engendered and complaints of false hope on the food supply to the ecolog-

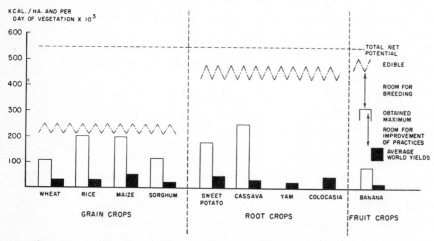

Figure 18–4   Current and potential food production for selected crops. (From Poleman, T.: Science, *188*:517, 1975.)

ical and genetic shortcomings of intensive agriculture. That is, recent criticism extends to the issues of increased erosion of farmlands everywhere and crop monoculture as a practice vulnerable to disease epidemics (and possible pesticide overdosing), as already discussed for corn blight. Also there is the related issue of a precariously narrow gene pool for the major crops of the world. When "miracle" wheat or rice is planted in poor countries, the locally adapted, traditional varieties are often discarded. These local varieties are an irreplaceable treasure of genetic diversity for disease-resistance and other attributes which could be useful in future breeding programs. We need to develop vigorous conservation programs for these local varieties; some seed banks have been set up, but much more needs to be done in this regard.

The final shortcoming of the green revolution to be mentioned here is its slowness in developing good nutritional qualities of the miracle grains. The first high-yielding varieties were not selected for enhanced nutritional value. Belatedly, the protein content of these grains is being looked at and minor improvement in content has now occurred. However, grains like rice, wheat, and corn have far lower protein content and protein quality than do peas, beans, lentils, and other legumes. And legumes have been abandoned on many a farm in Asia and elsewhere that switched to planting the new grains. Unfortunately, the green revolution leadership in the international breeding programs that now exist has not yet made a major effort to produce high-yielding peas and beans and to introduce them into poor countries. (Peas and beans have the additional advantage over grains of not needing expensive fertilizer because they are capable of "fixing" nitrogen in specialized structures of their roots.)

PLANNING FOR THE FUTURE

How might genetic, ecological, and production security be achieved for the average farmer of the world (who typically has an eight-acre field in a poor country)? There are two basic methods of attack on this problem. (1) Genetic studies of crops should be considered adequate only when the plant genotypes are understood in relation to the genotypes of possible pests and parasites, and in relation to climatic extremes and competing species such as weeds. In this ecological-genetic approach, suggested by Richard Levins (1974), a corn or wheat field is not simplistically regarded as a starch factory, but as a complex ecological system with many variables. (2) Plant geneticists should advertise and advocate not a monocrop farm plan (one-type seed plus fertilizer plus machinery), but a dependable multicrop plan characterized by medium yield overall but a low potential for total loss of food products. Although planting a mixture of crops is a more labor-intensive process, it makes it impossible for a single plant disease to destroy one's livelihood and drive one off the land. In the nonindustrialized countries, peasant farmers are better off at home than in faceless, jobless cities. In fact, even in the U.S. urban unemployment and rising energy costs will soon force increased manpower-intensive farming practices.

Other methods of improvement in food production, more sophisticated than the two described above, are now available (see discussion by Wittwer, 1975).

One historical note: improved medical practice during the twentieth century (including the recent advances in saving individuals from genetic conditions that were previously fatal) has resulted in unprecedented population expansion. In the 1950's and 60's the medical profession was rightly criticized for not fostering birth control programs with the zeal with which it fostered "miracle" cures with antibiotics. Now in the 1970's we have the agricultural planning profession still advocating "miracle" food with more zeal than is advocated for birth control programs. Both the medics and the agriculturalists have been highly motivated and well-intentioned. But partly because of the compartmentalization in human (and scientific) affairs, the agriculturalists did not perceive until very recently that they had repeated the former mistakes of the medics. Overspecialization is a "maladaptive trait" in this chapter of social history.

PLANT DOMESTICATION

All our crop plants in the world were once wild species. Most of them became cultivated before recorded history. We do not know how they became cultivated, but it is possible to imagine that the plant foraging activities of ancient hunting-gathering groups led easily to cultivation. Because seeds are often indigestible, eating plant parts that contained them would likely result in their becoming strewn around the living site. If the seeds were enclosed in fecal material they would be provided with a ready source of fertilizer with which to begin germination and growth. If the living site was still occupied the following year, edible plants would have been noticed to have sprung up close by, and would naturally have been utilized. Human practicality probably accounts for the deliberate strewing of seeds in select sites for the sake of convenience and for providing excess for storage for the off season. An added incentive for starting the deliberate cultivation of plants may have been the usefulness of surplus food for trade.

Until the 1920's there was no systematic knowledge of where plants important for human welfare originated. At that time a Russian by the name of Vavilov began traversing many areas of Asia and the Near East and collecting locally grown varieties of plants, as well as their uncultivated wild and semi-wild relatives. Vavilov reasoned that an area which contained great diversity of cultivated and uncultivated forms of a crop plant, and had ancient traditions for its use, was the center of origin— or gene center—for the plant in question. (We now know that there are both primary and secondary centers for many plants, but the important point is that areas do exist with great genetic diversity for useful types of plants.)

By his meticulous investigations Vavilov determined that the Cilician plain of Turkey was where flax, from which linen is made, originated and diversified. He and later investigators showed that such important crops as wheat, barley, chick peas, and lentils all originated in the Near East. More recent study has shown that sub-Saharan Africa gave rise to the sorghum family, the millets, and some cotton species and tropical fruit trees. The New World is the home of peanuts, cocoa, chilis, corn, tomatoes, some types of cotton, and potatoes—as well as turkeys and llamas. Other areas gave rise to other commercially important species.

Plant geneticists look for disease-resistant varieties of a plant in its home area, if it has not already been decimated there by modern changes: where the plant flourished and diversified it must have acquired—through mutation, recombination, and selection—a good defense against many kinds of pathogens and pest attacks. The subject of conserving useful genes is discussed in the next section.

Where did agriculture start? It started in many places and at many times. Using sophisticated methods of dating the age of seeds found at archeological sites, we can say that seeds that look very similar to conventional crop seeds have been found at human sites which are 7000 to 9000 years old throughout the tropics and subtropics of the Old World (Fig. 18–5). The earliest cultivated forms found are for rice,

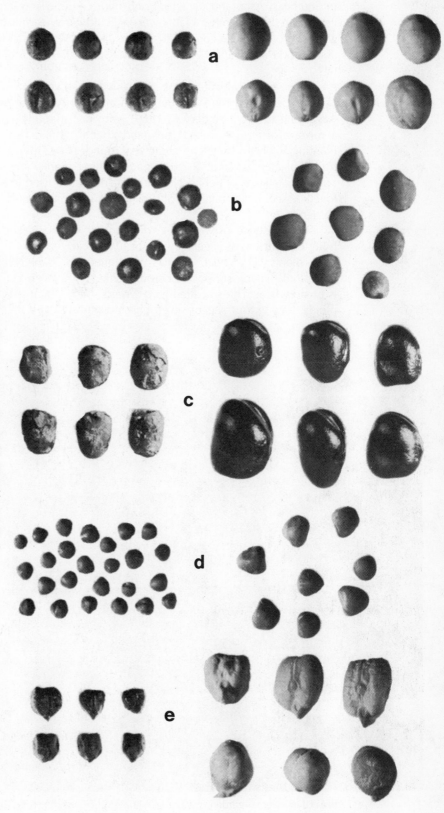

Figure 18–5 Seed remains (left) and, for comparison, seed from recent varieties (right) of the five cultivated pulses. (a) Pea, carbonized seed from Early Bronze Age Arad, Israel. (b) Lentil, carbonized remains from Late Bronze Age Manole, Bulgaria. (c) Broad bean, carbonized seed from Copper Age Chibanes, Portugal. (d) Bitter vetch, remains from Late Bronze Age Manole, Bulgaria. (e) Chickpea, carbonized seed from Early Bronze Age Arad, Israel. (Magnification about 1.8.) (From Zohary, D., and Hopf, M.: Domestication of pulses in the Old World. Science, *182*:887–894, 1973.)

in Southeast Asia, and for wheat and barley in Iraq and Iran, although some plant geneticists believe other plants were in cultivation earlier (e.g., squashes and sorghum). Fruit trees, according to studies in the Near East, were not cultivated until about 5000 years ago. The original cultivations were for dates, olives, grapes, and figs.

Many early cultivated plots probably looked very much like the mixed gardens of rural peoples throughout the warm areas of Asia and Africa today (Fig. 18–6). According to Edgar Anderson (1960), who has studied such gardens, they contain a mixture of trees, shrubs, vines, and annual plants. The plants are not in rows. Planting and harvesting tend to run year round. The crops from such a garden include fruits, vegetables, starchy roots, gourds, fiber plants, stimulants and other drugs, condiments such as pepper, cosmetics, and plants used in ritual—often all planted together seemingly haphazardly. Weeds are not ruthlessly excluded from such a garden.

The presence of weeds in contemporary peasant gardens suggests the possibility of their having been important in the early stages of domestication of small-grain plants like wheat. As was mentioned in Chapter Six, common wheat is a hexaploid, having six sets of chromosomes with seven chromosomes per set. This is bread wheat. The so-called hard wheats, from which spaghetti is made, are tetraploids, having four sets of seven chromosomes each. Other wheats, or wheat-like

Figure 18–6   Semai garden in Pahang, Malaysia, showing mixed cultivation.

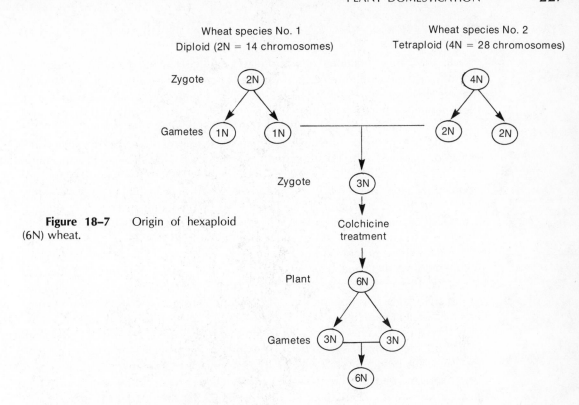

Wheat species No. 1
Diploid (2N = 14 chromosomes)

Wheat species No. 2
Tetraploid (4N = 28 chromosomes)

Zygote 2N

4N

Gametes 1N 1N

2N 2N

Zygote 3N

Colchicine
treatment

Plant 6N

Gametes 3N 3N

6N

Figure 18–7   Origin of hexaploid
(6N) wheat.

weeds as we might classify them, are tetraploid or diploid. These weeds contain
the key to the change wheat has undergone over the last three or four millenia.
There is suggestive evidence that in early times, and even to this day in home areas
of wheat, a mixture of wheat-like seeds was sown and grown together in a plot.
There were probably also reaped, harvested, and threshed and baked as a mixture.
Although some of these plants would be classified as different species, because they
hybridize only rarely, such a multitude of genotypes growing together in one cultivated plot certainly provides maximal opportunity for genetic recombination to
occur. From such a "game plan" it is easy to see that cultivated bread wheat did
not have a single origin. Indeed, we have strong evidence for its having a multiple
origin. By experimentally hybridizing diploid and tetraploid species of wheat-like
plants, geneticists have produced an instant hexaploid plant with features very similar to that of "natural" bread wheat. The ploidy situation in this experiment is as
follows: 1N gametes of diploids combine with 2N gametes of tetraploids to produce
a 3N (triploid) zygote; such a zygote is then artificially induced to become 6N by
application of an inhibitor of mitotic spindle formation, such as the drug colchicine
(Fig. 18–7).

For some other crops, in addition to wheat, the wild ancestral species that gave
rise to the modern crop form are still in existence and have been identified. This
is true for peas and lentils, for instance, which are known to have accompanied
wheat and barley in the Stone Age agricultural complex developed in the fertile
crescent area of the Near East. Nutritionally, this combination of grains and legumes
makes sense because the grains furnish the carbohydrates while the legumes provide
the necessary protein for a balanced vegetarian diet. A similar combination of
domestication occurred in ancient American agriculture (corn and beans).

PLANT GENE CONSERVATION

In 1967 an international group of plant geneticists met and loudly sounded the alarm over the dwindling supply of varieties of crop species and of the wild species that are their weedy relatives. Until then, many geneticists had glibly assumed that the seemingly inexhaustible range of plant species and varieties that had been around for thousands of years would continue forever (Table 18–3).

Wild species are doomed by the relentless clearing of primeval forests, the cultivation of grasslands, the irrigation and tillage of deserts, the sprawl of urbanization, and the widespread killing of weeds by herbicides. Primitive crop varieties are doomed by super varieties. In short, man the destroyer of natural and peasant habitats is diminishing the natural diversity of the world, including that diversity which is a treasury of local adaptations for success in producing crops. This erosion of our genetic resources may catastrophically affect the future.

But could we not use artificial mutagenesis on garden-variety plants and induce all the old "weedy" variability that existed before the advent of intensive agriculture? Theoretically, yes. But hunting "in the haystack" for the good mutant, or a coadapted group of mutants, is an impractical task. It is much saner to have tried-and-true genotypes of ancient varieties to work with in complex plant breeding programs.

The present policy of preserving the ancestral genetic base from which our important crops arose is effected by setting up "gene banks" through seed collections. This program was conceived largely by O. H. Frankel of Australia and is now being carried out both through international organizations and within such prosperous nations as the U.S. and the U.S.S.R. But the program is underfunded and understaffed. The erosion of plant genetic variability in the world increases every year. Continued neglect of this erosion will become more dangerous as time goes on. Before 1940 genetic erosion was already severe in the countries where plant breeding programs had long been active: the U.S., Canada, Japan, Australia, and New Zealand. And more high-yielding, green-revolution crops continue to be

TABLE 18–3   Time Scale of Concern About Plants*

Period	Type of Plant	Human Agent	Objective	Time Scale of Concern
To 8000 BC	Wild	Hunter-gatherer	The next meal	1 Day
8000 BC to AD 1850	Domesticated or semi-wild	Peasant farmer	The next crop	1 year
1850 to present	Domesticated	Plant breeder	The next variety	10 years
1900 to present	Domesticated + wild relatives	Crop evolutionist	To broaden the genetic base of crops	100 years
1960 to present	Domesticated + wild	Genetic conservationist	To maintain plant diversity on world scale	1000 years?

*Adapted from Table 1, O. H. Frankel (1975).

planted, to the exclusion of native varieties, in the poor countries of the world, many of which are important "natural gene centers." Already by 1948 native flax varieties were extinct in Turkey (the center of origin of flax), and only about 10 per cent of the rice-growing region of Africa is now planted to native rice strains.

A more expensive but probably more enriching system for conserving crop-related genes is to set aside "genetic resource areas" in Vavilov-type geographic centers. These would be living museums of farm areas with potentially useful genetic diversity where agriculture would be restricted to traditional ways so that weedy relatives of crop plants could exist in uncultivated zones and field plants could hybridize freely with wild relatives nearby. This idea, admittedly, seems an economic and social dream, considering the prevalence of wars and the fluctuating fortunes of governments. Seed banks seem a safer bet.

GENETIC ENGINEERING ON PLANTS

Among the many imaginative schemes suggested for greater world food production, two that turn on genetic research are worth mentioning. First is the engineering of synthetic hybrids between crop species that, hopefully, combine advantages of both parental types. One such synthetic but fertile hybrid already developed is triticale (from wheat × rye). Triticale may eventually compete well with other cereal grains in agriculture since it is better adapted to marginal lands than wheat and has the good protein quality of rye. About a half million hectares (over a million acres) of triticale were planted worldwide in 1973.

The second genetics-related innovation scheme involves the engineering of haploid plants by growing pollen cells in tissue culture in the laboratory. Haploid plants would produce only genetically uniform, true-breeding individuals. Such test tube breeding of haploid plants would allow for rapid screening of desirable mutations which were induced artificially, and, perhaps, for the production of novel species of useful plants which cannot be formed by standard sexual hybridization. For example, novel diploid cells can already be formed in the laboratory by fusing together haploid cells from different species. The ideal plant in the mind of some would be a cereal like rice or wheat which has the ability of a legume to produce its own nitrogen supply without the need of fertilizer. A rice-bean synthetic hybrid would relieve world agriculture of much of its fertilizer problems.

Most of this test tube research, however, is in its infancy. It exists as a hope on the horizon, not a promise.

SUMMARY

Practical genetic problems at the level of molecules, cells, and populations of individuals abound in the realm of agriculture. Human selection of farm species has advanced from the traditional method of selecting for vigorous phenotypes to selecting for a variety of cryptic attributes, including disease resistance and protein quality. Still, given the somber prospects of human population increase, the results of agrogenetic research have been disappointing. There is only moderate hope for substantial improvement of the world food supply in the future.

STUDY QUESTIONS

1. Farmers buy new hybrid corn seed for planting every year, rather than saving seed ears from their field for sowing the subsequent year. Suggest a genetic reason for this.

2. Onions come in three colors: red, yellow, and white. Reds have the genotype C– R–, yellows have C– rr, and whites have cc rr or cc R–. C/c and R/r are independently segregating genes. C and R permit the synthesis of a plant "antibiotic" in conjunction with the synthesis of red and yellow pigments. The antibiotic confers resistance to mold diseases. What is the likely action of cc at the level of plant metabolism?

3. Discuss the advantages and disadvantages of Americans becoming vegetarians.

4. Key terms:

   hybrid corn   green revolution   genetic engineering
   monoculture   gene conservation

REFERENCES*

Anderson, E. The evolution of domestication. *In* Evolution After Darwin, Vol. 2, S. Tax (ed.). Chicago: University of Chicago Press, 1960, pp. 67–84.

Frankel, O. H. Genetic conservation: our evolutionary responsibility. Genetics *78*:53–65, 1975.

Gavan, J. D., and J. A. Dixon. India: a perspective on the food situation. Science *188*:541–549, 1975.

Levins, R. Genetics and hunger. Genetics *78*:67–76, 1974.

Mangelsdorf, P. C. Corn: Its Origin, Evolution and Improvement. Cambridge, Mass.: Harvard University Press, 1974.

Poleman, T. T. World food: a perspective. Science *188*:510–518, 1975.

Wittwer, S. H. Food production: technology and the resource base. Science *188*:579–584, 1975.

SUGGESTED READINGS

Brown, J. W. Native American contributions to science, engineering, and medicine. Science *189*:38–40, 1975.

Carlson, P. S., and J. C. Polacco. Plant cell cultures: genetic aspects of plant improvement. Science *188*:622–625, 1975.

Chang, K-C. The beginnings of agriculture in the Far East. Antiquity *44*:175–185, 1970.

Cox, G. W., and M. D. Atkins. Agricultural ecology. Bulletin of the Ecological Society of America *56(4)*:2–6, 1975.

Curtis, B. C., and D. R. Johnston. Hybrid wheat. Scientific American *220*:21–29, 1969.

Day, P. R. Genetics of Host-Parasite Interaction. San Francisco: W. H. Freeman & Co., 1974.

Flannery, K. V. The ecology of early food production in Mesopotamia. Science *147*:1247–1256, 1965.

Greenland, D. J. Bringing the green revolution to the shifting cultivator. Science *190*:841–845, 1975.

Harlan, J. R. Our vanishing genetic resources. Science *188*:618–621, 1975.

Harris, M. The withering green revolution. Natural History, March, 1973. pp. 20–23.

Heiser, C. B. Seed to Civilization: The Story of Man's Food. San Francisco: W. H. Freeman & Co., 1973.

---

*Note: items cited in the References are also recommended for suggested reading.

Hightower, J. Hard Tomatoes, Hard Times. Washington, D.C.: Agribusiness Accountability Project, 1972.
  Criticizes American agriculture for breeding crops for machine-handling convenience rather than with the consumers' interests upmost.
Hulse, J. H., and D. Spurgeon. Triticale. Scientific American 231:72–80, 1974.
Mayer, A., and J. Mayer. Agriculture, the island empire. Daedalus 103(3):83–95, 1974.
Miller, J. Genetic erosion: crop plants threatened by government neglect. Science 182:1231–1233, 1973.
Reitz, L. P. New wheats and social progress. Science 169:952–955, 1970.
Staub, W. J., and M. G. Blase. Genetic technology and agricultural development. Science 173:119–123, 1971.
Strickberger. M. Genetics. Ed. 2. New York: Macmillan Publishing Co., Inc., 1976.
  Chapter 33 gives an excellent description of hybrid corn production.
Tatum, L. A. The southern corn leaf blight epidemic. Science 171:1113–1116, 1971.
Ugent, D. The potato. Science 170:1161–1166, 1970.
Wade, N. Green revolution: creators still quite hopeful on world food. Science 185:844–845, 1974.
Wade, N. Green revolution. I [and] II. Science 186:1093–1096 and 1186–1192, 1974.
Walters, H. Difficult issues underlying food problems. Science 188:524–530, 1975.
Williams, P. H. Genetics of resistance in plants. Genetics 79:409–419, 1975.
Zohary, D., and M. Hopf. Domestication of pulses [legumes] in the Old World. Science 182:887–894, 1973.

It might be supposed that the study of the evolution of behavior would have little to contribute to understanding new social institutions, but the study of evolution tells us a great deal about how animals learn. . . . Early experience is tremendously important, particularly so in the case of slowly maturing man. Learning first takes place in emotional, close interpersonal situations and is later motivated by clearly defined objectives. Repetition is guaranteed by pleasurable activity until skills are mastered. Peers are important in transmitting education. Rewards are highly specific. The traditional American educational system has not been operated in accord with what we know about how primates learn. It starts too late, relies on words rather than actions and on discipline rather than play. The goals are not clear to either teacher or student, and peers are excluded from the role of teacher. It would be hard to devise a worse system for educating human beings. The study of how humans learn and of the evolution of learning is very different from the study of schools. The study of schools cannot possibly tell us how humans learn because natural learning situations have been excluded from the activities in the schools.

S. L. Washburn and E. R. McCown (1972)

Chapter Nineteen

HUMAN EXISTENCE: MAINTAINING HUMAN DIVERSITY

It is important that genetic diversity be maintained among the people of the world so that genetic adaptations to local environments are preserved and so that alleles crucial to survival under future conditions will be in existence when new conditions are encountered. Human genetic diversity is also desirable because, as one wag put it, "uniformity among people is boring."

Before we consider the maintenance of genetic diversity from this point onward in human existence, let us review what is known about the development of the human stock in the past.

PAST HUMAN EVOLUTION

Human beings are grouped among the primates, a classification of mammals that also includes monkeys and apes (chimpanzee, gorilla, orangutan, and gibbon). The term human evolution often is used to encompass both human and prehuman evolution, the latter being the events inferred to have led up to the appearance of *Homo sapiens*. Our fragmentary knowledge of prehuman evolution rests almost entirely on (1) a comparison between living humans and higher primates such as the chimpanzee and gorilla, and (2) a comparison between fossil bones of early hominids (human-like forms) and bones of present apes and humans.

Our closest living relative is the chimpanzee. Human hemoglobin and a variety of other proteins such as cytochrome C and fibrinopeptides A and B are identical to those of chimpanzees. Overall, living apes are most like humans in body chemistry (human and chimp proteins are probably 99% identical, according to M. C. King [1975]), and less like humans in anatomy and behavior. Anatomically, the chimpanzee has changed little in the last five million years, whereas the human line has changed dramatically in anatomy, and, concomitantly, in behavior. Indeed, the significant distinguishing characteristics of the human species (Fig. 19–1) are behavioral: great learning ability, language, and culture—that is, major intentional modification of the physical and social environment. Tools are a means of modifying the physical environment. Several investigators have shown that chimpanzees

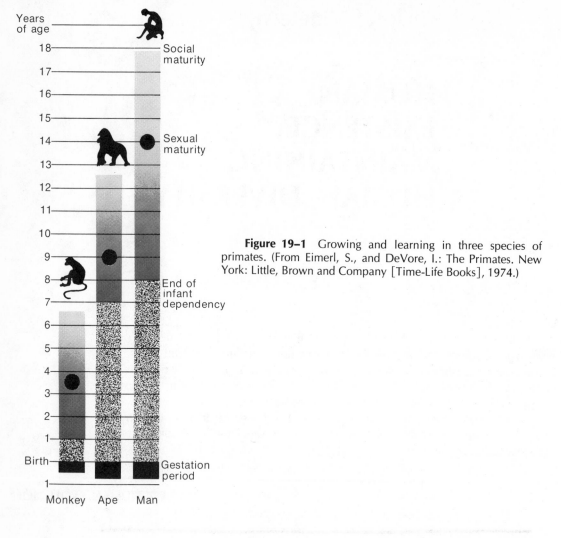

Figure 19–1 Growing and learning in three species of primates. (From Eimerl, S., and DeVore, I.: The Primates. New York: Little, Brown and Company [Time-Life Books], 1974.)

and other primates are tool-users (Fig. 19–2), though not so habitually as early hominids and humans—with their penchant for chipping stone into useful shapes—are thought to have been. (See Box on Human Characteristics, p. 235).

Modifications of the social environment include rituals, rule formation, burial of the dead, and a host of other factors. Recently an analysis of the plant remains in the soil surrounding a cave burial in Iraq about 60,000 years ago revealed that the corpse was buried with flowers. Evidence for the flower burial consists of flower-like concentrations of preserved pollen. Eight species of flowers, including the cornflower, grape hyacinth, and hollyhock were identified in the burial site. This discovery evokes an image of peculiarly human grief and esthetic perception over an immense time.

The fossil hominid finds now on hand are mainly from Africa. Future work in Asia and elsewhere may greatly modify our African perspective, however. Many sizes of hominid bones have been found throughout southern and eastern Africa, sometimes in sediments of the same age from the early Pleistocene period (2 to 3 million years ago). These findings are often interpreted as an overlapping succession of different hominid species (species of small-brained *Australopitheces* as well as *Homo*). A minority view interprets the findings as possible evidence of a single

Figure 19–2   A tool-using chimpanzee. (From Eimerl, S., and DeVore, I.: The Primates. New York: Little, Brown and Company [Time-Life Books], 1974.)

species which contained a very large range of adult sizes (from 3 to 6 ft. in height, reportedly). Living humans today, from pygmy to basketball player, certainly differ widely in adult size.

HUMAN CHARACTERISTICS

Homo sapiens is distinguished by upright posture and bipedalism, characteristics dating back over three million years, and a large brain relative to body size—the increase in brain size having begun about two million years ago. Food sharing, large-animal meat eating, tool manufacture, and the use of a home base are also thought to have begun about two million years ago. Other distinguishing characteristics include a lengthy childhood, extensive learning ability, language, and flexible social groupings.

When were prehumans alive? From at least 3.3 million years ago to one million years ago in Africa (Fig. 19–3). The ancestors of both humans and chimpanzees are thought to have diverged five or more million years ago. If there was more than one early hominid gene pool, only one gave rise to humans. The other or others died out. That is, our species is unusual, but not unique, in that it is the only living member of its group (the genus *Homo*). If other *Homo* species died out, the cause may have been poor competition with *H. sapiens* or its predecessor for food and other resources. Other plausible causes of species extinction, according to evolutionary theory, include predation (being eaten by some carnivore, but not necessarily a human one), parasitism (an unknown epidemic?), and inability to survive or move away from an environmental stress (ice age?). As Dobzhansky (1972) has written, "Evolution is a synthesis of determination and chance, and this synthesis makes it a creative process. Any creative process involves, however, a risk of failure, which in biological evolution means extinction" (p. 174).

If other *Homo* species became extinct, why then has the chimpanzee, with its relatively conservative anatomy and brain, survived? The chimp, as well as other apes, is a forest dweller and an eater of fruits and leaves. Chimps now occupy a smaller range than formerly and may, in fact, be near extinction as the result of being outcompeted by early humans over a long period. Alternatively it is possible, according to one popular theory, that early chimps did not compete with early humans in Africa for a habitat or food. At the time of human emergence the forests were shrinking and the grasslands expanding as the result of climatic changes. Early humans lived in open country and sought their food there. The chimp may have been unable to exploit this new habitat and thus became restricted in its range as the forests retreated.

From a genetic perspective we can suggest some circumstances involved in human emergence. On the basis of the human characteristics that appeared about two million years ago in this open-country habitat (see Box), it appears that selection favored a complex of social traits (food sharing, tools, home base), good nutrition (meat eating), and braininess. This complex of human traits provided a new formula for success in the biological world.

The increase in brain size in the human line (Fig. 19–3) is one of the most rapid evolutionary changes ever documented. What selective pressures of the environment favored this human characteristic? We can only guess at the processes involved. But it is well to bear in mind the lengthy history of the genus *Homo*: two million years. If a generation averaged 20 years over this period, this history represents a span of 100,000 generations. (For comparison, the female chimp is sexually mature at nine years of age.) Many mutations could have arisen in that many generations and could have been selectively enriched in the human gene pool. Through genetic recombination, many new combinations of alleles could have been tested by the environment.

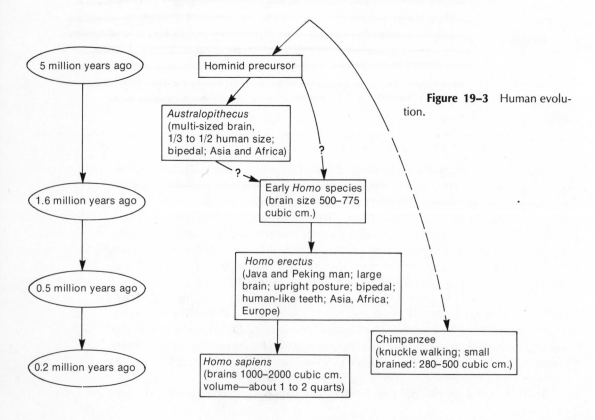

Figure 19–3 Human evolution.

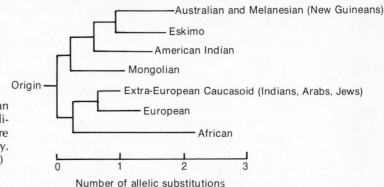

Figure 19–4 Tree of human evolution. (Modified from Cavalli-Sforza, L.: Population structure and human evolution. Proc. Roy. Soc. Biol., *164*:362–379, 1966.)

However, the major adaptive changes in the human line were seemingly not caused by a rapid rate of protein evolution, that is, by mutations in structural genes. (Recall that human and chimp proteins are almost identical.) M. C. King (1975) and others have suggested that the adaptive changes in the human line are the result of (1) mutations in critical regulatory genes, such as promotors or operators (see Chap. Nine), which altered the production but not the structure of enzymes, or (2) mutations in the order of genes on chromosomes, possibly affecting the timing of gene expression during human development (Fig. 19–5). Much more information is obviously needed on mutation rates and on the control of gene expression before we can delve deeper into the genetic mechanisms operative in human evolution.

EARLY HUMANS VERSUS MODERN TIMES

Early *Homo sapiens* were scattered throughout Africa, Asia, and Europe. They are categorized by the complexity of their stone tools. Early human groups were undoubtedly small, semi-isolated breeding populations which adapted both genetically and culturally to local conditions.

At the end of the Pleistocene period (the last of the ice ages) about 15,000 years ago, some of the lithic (stone-using) cultures turned gradually from hunting, fishing, and gathering to agriculture. A genetic perspective on this agricultural revolution was given in the previous chapter. For Europeans the agricultural center was the Near East (Mesopotamia). In addition to many plants, animals such as goats, cattle, sheep, and pigs were domesticated in the Near East starting around 10,000 years ago. From the Near East farming spread to Europe, reaching southern France 6000 years ago and Britain and Scandinavia 5000 years ago.

From a genetic point of view, these historical facts are of some interest. The human geneticist Cavalli-Sforza (1974) suggests that the domestication of milking animals such as cattle and goats allowed an increase in the frequency of the previously rare allele for lactose tolerance in milk-drinking populations. Lactose is milk sugar; it doesn't occur in other foods. Pre-weaned children are tolerant of lactose, but the majority of older humans throughout the world are intolerant of lactose. They lack the enzyme lactase in their intestinal tract and are thus unable to break down lactose into smaller, usable sugars for nutrition. Intolerant individuals experience diarrhea and discomfort when they drink milk. Post-weaning lactose tolerance and intolerance are thought to be genetic characteristics. Caucasians, traditionally a milk-drinking population, are largely lactose tolerant, but most other groups are lactose intolerant. Lactose tolerance allowed Caucasians to exploit an available food resource of their cultural environment. Lactose tolerance was evolutionarily adaptive in Europe. However, cows providing milk are also available in non-European areas such as India: there the intolerant population has largely circumvented its hereditary lactase deficiency by utilizing milk nutritionally in a fermented form such as yogurt. In yogurt the lactose content has been greatly reduced by its conversion to simpler sugars (by microorganisms which *do* have lactase).

It was only 200 or so years ago that some of the traditional agricultural cultures of the world turned into industrial-technological complexes and that science came into its own. During the entire span of human history the population has doubled and redoubled so that today almost four billion people exist, whereas a few hundred thousand at most existed 200,000 years ago when *Homo sapiens* began. The cause of this population increase is commonly said to be the increased efficiency of resource utilization occasioned by technological "breakthroughs" (e.g., plant domestication for traditional agriculture and the use of fertilizers, tractors, irrigation, etc., to provide food for industrial states). Ecologists say that such breakthroughs raise the "carrying capacity" of the environment—that they allow the environment to sustain, or carry, an increased population.

Having referred to the changes from a hunting-gathering population to an agricultural and then to a technological society, we come to the next important question: does evolution favor an increase in the organizational complexity of a population? (Similarly, one may question whether evolution favors increased complexity in an organisms's anatomy.) A number of biologists have argued that, contrary to first impressions, there is no inherent tendency in evolutionary processes toward complexity, toward progression or progress in evolution. We know of no component of natural selection, or of mutation or recombination, that generates complexity. Progress is neither inevitable nor intrinsically desirable. Consider the increased societal complexity found in technological states: today there is a nagging fear that modern life might be overly complex to the point that one sector may, through its specialization of interests, cause the demise of all—through atomic warfare, noxious pollution, or some other means. That is to say, although technology may have been adaptive for a while, having contributed to increased survival and fertility in the population, it may become evolutionarily maladaptive in the near future. It is at least within the realm of possibility that previously adaptive changes could ultimately cause species extinction.

There is another aspect of this propensity of humankind to increase at a rapid rate. All major societies are overproducing people, that is, none of them has yet accomplished zero population growth. Thus Garrett Hardin (1972) has drawn our attention to the possible problem of individuals within a society who voluntarily control their reproduction having more "unselfish" genes than those members of the population who reproduce lavishly. Hardin suggests that, if selfish and unselfish genes exist, conscientious people will be outbred by unconscientious ones. He has urged the consideration of compulsory laws to enforce low reproduction. Among Hardin's many commentators, Kirk (1972) argues that people reproduce at the level expected by their culture and points out that in the U.S. the birth rate has been declining since 1957, before "the pill." This may reflect a growing realization that children are economic liabilities in our present culture patterns. Kirk also argues that with contraception and abortion now increasingly available, perhaps a time will come when only people who truly want to have children will be having them. And if there are genes for good parenting, they should be thereby enriched in the population. (There are many other facets to this controversy; the interested reader should consult the writings by Hardin, Kirk, and Miles listed at the end of the chapter.)

RACE, CULTURE, AND GENES

Despite the evident cultural diffusion of ideas and the gene flow between human populations over the contours of the earth during the last 200,000 years or so, we still see today distinct differences among cultures and what we are prone to call races. A race, biologically defined, is a subdivision of a species having certain shared biological traits which set that group apart from other groups. For humans, the race-identifying group of traits is traditionally considered to be skin color, facial characteristics, and body build (but not, for example, susceptibility to malaria or lactose intolerance—why?). Human races are thought to have formed by the fragmentation of the human population into relatively isolated groups, all living in somewhat different environments; genetic drift and local adaptation via natural selection would then have brought about divergence among the groups.

To understand the origins of and relationships between human races, the genetic similarities and differences among them have been studied. The extent of genetic differences among races is often illustrated by the frequencies of polymorphic alleles in different groups. Genetic polymorphisms provide us with a large number of traits of many different types for analysis and they are less at the mercy of short-term environmental fluctuations than are such physiological traits as body weight, head size, and height. Many polymorphisms are known to show differences in allele frequencies among various areas of the world (see Table 17–3).

A representative study of genetic differences among races is that of Cavalli-Sforza in 1966. He studied the frequency distribution of 38 alleles for various immunological and blood-serum factors in a sample of human populations from throughout the world. By mathematical analysis of the allele frequencies he was able to construct a "phylogenetic tree" of genetic distances among major recognized human races (Fig. 19–4). The greatest genetic distance was found between Africans and Eskimos (4.6 arbitrary units) and the smallest between Europeans and extra-European Caucasoids (1.1 units). The Mongoloid group (top half of Fig. 19–4) is the most heterogeneous, suggesting that it fragmented at a relatively early time in human existence.

Turning now to a consideration of culture: a culture is a set of beliefs, norms of behavior, and customs. Most people of a culture also look alike, which is somewhat like saying that members of a race often have similar cultures. No race, however, is completely isolated genetically or culturally from other races. There are probably many interactions between biological traits and cultural ones. For example, Wiesenfeld (1967) has conjectured that the clearing of jungle in Africa for early agricultural settlements may have permitted the formerly jungle-roving mosquito vectors of human malaria to congregate successfully in the settlements where a supply of victim's blood would be continuously available. Any previous dispersal of humans throughout the jungle would probably have permitted only rare encounters between humans and malaria-bearing mosquitoes, which would then survive by attacking a variety of warm-blooded animals. The congregation of both people and mosquitoes in jungle clearings forced a sufficient rise in the frequency of the sickle-cell allele to permit survival of the agriculturalists. Supporting this idea is the fact that the Pygmies in the undisturbed African jungle today have low frequencies of the sickling allele.

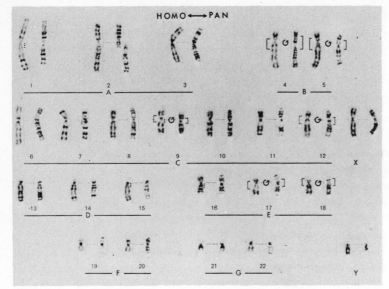

Figure 19–5 Comparison of karyotypes of man (left) and chimpanzee (right). The chimp chromosomes were arranged to show maximum homology with the human chromosomes. (From Sutton, H.: An Introduction to Human Genetics, 2nd ed. New York, Holt, Rinehart and Winston, 1975.)

Another example of the interaction between social conditions and genetic traits involves a small group in southeast Asia (Malaysia). The Temuan are a jungle-fringe group subsisting mainly on gathering, fishing, and slash-and-burn agriculture. Their environment is malaria saturated. The Temuan are largely isolated by their belief systems and life style from neighboring groups of Chinese, Indians, and Malays. (Malaysia is a pluralistic society.) The strong genetic isolation of the Temuan is advantageous to them in their traditional habitat because their gene pool contains several malaria-resistant genes in moderate frequency: a hemoglobin variant, G6PD deficiency (Chap. Nine), and a variant of red blood cell shape called elliptocytosis. The groups neighboring on the Temuan have these malaria-resistance factors only in negligible frequency. The cultural identity of the Temuan, that is, their ethnocentricity, fosters the maintenance of their complex of red-cell traits adapted to malarial resistance.

It is curious that ethnocentricity is a characteristic of every identifiable racial or cultural group. Whatever the social harms of ethnocentricity, or racial prejudice — and they are many — ethnocentricity does perform the biological function of retaining adaptive gene complexes within a defined gene pool. This biological implication of ethnocentricity has only recently become of interest to workers in human populations genetics. Explanations of human cultures based on biological adaptations may some day eventuate in a lively amalgamation of human genetics and anthropology.

POLITICS AND EUGENICS

Most people think politically and racially in a typological way. A Democrat or Italian is considered a static "type." When we say that all Democrats or Italians think or are alike, we are stereotyping. Dobzhansky (1968) has pointed out the problems of such simplifications; he urges us to substitute pluralistic, populational thinking in our social interactions in order to put us closer to the biological realities of the world.

One of the yearnings of humans in the Western world is seemingly toward some idea of perfection (which, however, varies among individuals and cultures). Plato talked about ideal types and prototypes. Today fashion magazines still talk about ideal types, and teenagers in particular are prone to find perfection in their idols. In the nineteenth century the yearning for perfection eventuated in the development of the eugenics movement in western Europe and the United States. The movement was started by Francis Galton, a cousin of Charles Darwin. Galton was interested in the genetic basis of "genius" and studied many well known families in England to that end, without, however, coming to any conclusions as persuasive as those of Mendel (about whose work Galton was ignorant).

Eugenics is concerned with improving the human organism, and gene pool, by weeding out physical and mental defects (negative eugenics) and by augmenting "desirable" traits (positive eugenics). Negative eugenics can be accomplished either by not permitting people to reproduce who have serious hereditary defects or whose offspring will be defective, or by the elimination of defective progeny by abortion or other measures. In Western society the prevention of offspring with a serious defect is partially accomplished by voluntary nonreproduction (see Chap. Fifteen) and involuntary sterility, such as occurs in segregated mental institutions. Although infanticide has been traditionally acceptable in some cultures, in technological cultures it is abortion, such as for fetuses with Down's syndrome, that has become sanctioned. Infanticide poses far more moral problems within the Judeo-Christian tradition. However, by far the greatest proportion of "negative eugenic" measures happen automatically: severely retarded persons, especially, are sterile by nature. Negative eugenics as a means of achieving perfection is now realized to be a vain hope because of handicapping mutations which recur spontaneously: most such recessive alleles are in the heterozygous state where they cannot be detected or selected against, and, at least in some instances, heterozygotic advantage exists for carriers of certain harmful mutations.

As Matsunaga (1966) in Japan and others have pointed out, the small size of families today in industrial countries—with most children born to mothers between the ages of 20 and 30 years—means minimization of Down's syndrome and other chromosomal abnormalities correlated with increased maternal age, of Rh and ABO incompatibilities between mother and fetus, and of some other genetic syndromes as well.

Positive eugenics is theoretically of wide scope: consider the vast array of domesticated varieties of chickens, dogs, cattle, and cats produced by deliberate selection of certain phenotypes. And now the choice of human phenotypes which might some day be selected for has been greatly expanded by the recognition of polymorphisms at the protein level that may be confidently assayed with techniques such as electrophoresis.

In recent decades Hermann Muller publicized his plan for voluntary artificial insemination of women by select sperm donors as a means of positive eugenics. Such a program has not been envisioned for eggs from select women because the technology of egg collection and their reinsertion into foster-pregnant women is complex, as recent work along these lines by R. G. Edwards has made clear (see also Chap. Twenty).

The problems associated with Muller's proposal have been hotly debated. They include, (1) the question of the healthiness of sperm samples stored for a long time (in a frozen state), (2) the question of the genetic effectiveness of any such program that is not mandatory, and (3) the question of which men are select—that is, have agreed-upon favorable traits. If, for example, only men of I.Q. scores of 122 or greater (six per cent of the male population) were allowed to reproduce, but all women were allowed to do so, according to Cavalli-Sforza and Bodmer (1971) the gain in I.Q. scores for the next generation would be about four per cent (increasing the average score from 100 to 104). Under such a selection regime, however, the gain would be less and less in each succeeding generation, eventually leveling off at an average I.Q. value near 130. Would this be a desirable trend? We know some geniuses are "difficult." And, so far as is known, a high I.Q. score does not produce an increase in human kindness. Indeed, it is possible that having more high-I.Q. people would bring about more technology for wars. On the other hand, some social planners have pointed out that more brains are always needed to engage in meaningful diplomacy, to create new energy-saving devices and more effective contraceptives, and to cure cancer. Overall, a mass experiment in positive eugenics would be an unhedged bet with fate.

Hitler and the Eugenics Movement

The largest, most notorious attempts at both positive and negative eugenics occurred in Germany between 1933 and 1945. They were put to an end with the end of World War II. Before we discuss this ignominious chapter in history, let us look at the development of the eugenics movement. Established in England by Galton in 1907, the movement began in the U.S. as the Eugenics Record Office in 1913. Other western European countries also started similar organizations, as did even Japan.

These movements were led by well-meaning, enthusiastic men who usually were not fully conversant with science in general or genetics in particular. Some did worthless or socially harmful research, although a few of their studies made useful additions to the growth of human genetics. They were overzealous in attacking feeblemindedness, criminality, insanity, pauperism, and so on—too eagerly attributing these conditions to a solely genetic cause. This is called the hereditarian position. Not surprisingly, it gave rise to an environmentalist reaction in the social sciences. Already in evidence at mid-century, environmentalism still dominates these disciplines today.

In the U.S. the hereditarian bias of the eugenicists gave rise to several political interventions: segregating the mentally defective in total-care institutions "for their own good," passing an enormous number of state sterilization laws for the feeble-minded and insane, passing state laws prohibiting marriage of the insane and retarded and marriages between races (particularly in the South), and passing federal laws restricting immigration largely to people from northern Europe, on the argument of their superior intelligence (see Chap. Fourteen). Many geneticists who had been indifferent to the eugenics movement (the chronic problem of the social unconcern of scientists), or who had even supported it initially, became critical in time as it became an activist success and as more became understood about human genetics and evolutionary principles: social ills did not have a simple genetic cause and the gene pool seemed to be stable, rather than "degenerating" at an appreciable rate. Moreover, by the 1930's geneticists generally were repelled by the pseudo-genetic racist politics of the Nazi regime in Germany.

The Nazis came to power as a result of economic difficulties in Germany generated by World War I and the Great Depression. Hitler chose to restore German self-confidence by myths of racial superiority and by scapegoating other races for somehow causing Germany's failures in war and peace.

The Nazis advanced their political position by the following policies:

1. Sterilization laws (in 1933): Persons with insanity, retardation, etc., were ordered sterilized by the "hereditary health courts" set up in every town. Over 350,000 sterilizations were performed. (In the U.S., at the height of the eugenics fervor, about 20,000 sterilizations were performed in 1935.)

2. Marriage prohibition laws: The Nuremberg laws of 1935 in Germany prohibited anyone from marrying a person with an infectious disease, mental disorder, or hereditary disease. Marriage between Jews and non-Jews was prohibited, as was marriage between half-Jews and non-Jews, but quarter-Jews could only marry non-Jews. The penalty for nonsanctioned marriages was imprisonment (Slater, 1936). (The U.S. also had marriage prohibition laws — as well as social prohibitions against interracial marriages.)

3. A bonus was given to German families of "good" phenotypes upon the birth of their third or subsequent child. (Although the U.S. had no such positive eugenics scheme, the ability of the rich to buy their way out of the draft in World War I and to obtain student deferments in later wars suggests a positive eugenics rationale.)

4. Breeding programs were established in Germany for producing "superior" offspring from parents exhaustively screened for medical and family defects. To this day very little has been published on the extent or outcome of these programs. Most likely the Nazi regime did not worry much about genetic recombination effects in setting up this scheme, and probably expected the offspring to be a blend of the traits of the two parents.

5. Genocide was carried out against Jews, Poles, mental retardates, and others, after 1938. These gas chamber murders were Hitler's "final solution" to the race purification program. The Nuremberg trials after World War II amply documented the extent of this pseudoscientific insanity: at least six million human beings were killed.

How did geneticists in Germany react to all these programs? Many of the German geneticists, some of whom were Jewish, emigrated to the U.S. before World War II broke out. A few German geneticists, however, agreed with and promulgated in writing the Nazi racist policies. Among these was von Verschuer, who applauded Nazi race purification during the war; in 1962, however, he repudiated it. In a discussion of this period, L. C. Dunn (1962) wrote:

> Although not all geneticists who remained in Germany accepted the eugenical and racial doctrines and practices of the Nazis, there is at least evidence that even the serious scientists among them underrated the dangers of the movement until it was too late. From this the melancholy historical lesson can be drawn that the social and political misuse to which genetics applied to man is peculiarly subject is influenced not only by those who support such misuse, but also by those who fail to point out, as teachers, the distinctions between true and false science. (p. 7.)

SUMMARY

Humans are descended from a chimp-like ape, by way of several intermediary species of *Australopitheces* and *Homo*. The cause of human emergence is unknown, but evolutionary theory provides plausible explanations for the adaptive advantage of general human traits. Molecular-genetic characteristics now known for humans are inadequate to explain human emergence: perhaps regulatory mutations will in the future provide an explanation.

The cultural diversification of humanity throughout the globe was accompanied by genetic diversification, at least in some instances by genetic adaptation to the local environment through selection favoring the survival of a particular variant allele. Exploitation of the environment everywhere for human ends has, among other things, produced overpopulation. Agricultural societies seem to be the worst perpetrators of overpopulation, on the basis of current evidence. An acceptable solution to overpopulation is not yet at hand.

Eugenicists urged the acceptance of methods to control population quality in the first half of the twentieth century in the Western world. The scientifically unsound eugenics movement eventuated in the despicable racist practices of Hitler in Germany, an object lesson on overzealousness for reformist policies whose scientific components are poorly understood and whose consequences have not been well thought through.

STUDY QUESTIONS

1. Discuss the possible selective factors operative in human evolution.
2. Discuss possible selective factors operative now on the human population.
3. The factors that cause genetic divergence in race formation are thought by Cavalli-Sforza and Bodmer to inevitably cause cultural divergence. Discuss.
4. Do you think it it likely today in the U.S. for eugenics enthusiasts to control human reproduction? Why, or why not?
5. Key terms:

primate	progress	eugenics
hominid	culture	hereditarian
lactose tolerance	race	race purification
lactase deficiency	racism	

REFERENCES*

Cavalli-Sforza, L. L. The genetics of human populations. Scientific American 231:81–89, 1974.

Cavalli-Sforza, L. L., and W. F. Bodmer. The Genetics of Human Populations. San Francisco: W. H. Freeman & Co., 1971.

Dobzhansky, T. Darwinian evolution and the problem of extraterrestrial life. Perspectives in Biology and Medicine 15:157–176, 1972.

Dunn, L. C. Cross currents in the history of human genetics. American Journal of Human Genetics 14:1–13, 1962.

Hardin, G. Exploring New Ethics for Survival: The Voyage of the Spaceship Beagle. New York: Viking Press, Inc., 1972.

King, M. C., and A. C. Wilson. Evolution at two levels in humans and chimpanzees. Science 188:107–116, 1975.

Kirk, D. Man's evolutionary future. Social Biology 19:362–366, 1972.

Matsunaga, E. Possible genetic consequences of family planning. Journal of the American Medical Association 198:533–540, 1966.

Slater, E. German eugenics in practice. Eugenics Review 27:285–295, 1936.

Washburn, S. L., and E. R. McCown. Evolution of human behavior. Social Biology 19:163–170, 1972.

Wiesenfeld, S. L. Sickle cell trait in human biological and cultural evolution. Science 157:1134–1140, 1967.

Note: items cited in the References are also recommended for suggested reading.

SUGGESTED READINGS

Baer, A. Lactase deficiency and yogurt. Social Biology *17*:143, 1970.

Baer, A., L. E. Lie-Injo, Q. B. Welch, and A. N. Lewis. Genetic factors and malaria in the Temuan. American Journal of Human Genetics *28*:179–188, 1976.

Birdsell, J. B. The problem of the evolution of human races: classification or cline? Social Biology *19*:136–162, 1972.

Cavalli-Sforza, L. L. Some current problems of human population genetics. American Journal of Human Genetics *25*:82–104, 1973.

Dumond, D. E. The limitation of human population: a natural history. Science *187*:713–721, 1975.

Eckhart, R. B. Population genetics and human origins. Scientific American *226*:94–103, 1972.

Eckland, B. K. Trends in the intensity and direction of natural selection. Social Biology *19*:215–223, 1972.

Eimerl, S., and I. DeVore. The Primates. Boston: Little, Brown & Co., 1974.

Goldsby, R. Race and Races. New York: Macmillan Publishing Co., Inc., 1973.

Kolata, G. B. Human evolution: life-styles and lineages of early hominids. Science *187*:940–942, 1975.

Kretchmer, N. Lactose and lactase. Scientific American *227*:70–78, 1972.

Ludmerer, K. M. Genetics and American Society: A Historical Appraisal. Baltimore: Johns Hopkins Press, 1972.

McHenry, H. M. Fossils and the mosaic nature of human evolution. Science *190*:425–431, 1975.

Miles, R. Whose baby is the population bomb? Population Bulletin *16*:3–36, 1970.

Oxnard, C. E. The place of the australopithecines in human evolution: grounds for doubt? Nature *258*:389–395, 1975.

Smith, J. M. Molecular evolution and the age of man. Nature *253*:497–498, 1975.

Solecki, R. S. Shanidar IV, a Neanderthal flower burial in northern Iraq. Science *190*:880–881, 1975.

Wolpoff, M. H., and C. L. Brace. Allometry and early hominids. Science *189*:61–63, 1975.

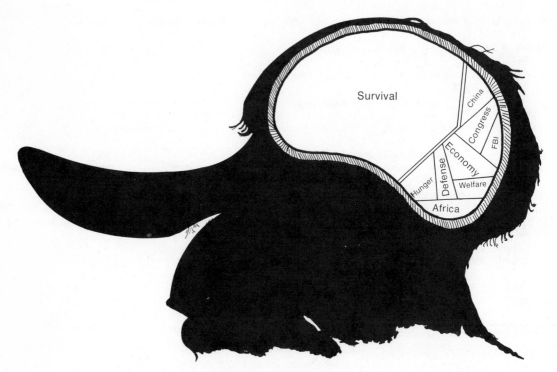

"Survival." Don Wright, Miami News. Distributed by Washington Star Syndicate. From: Graphics '75 Watergate. R. B. Freeman. Copyright © R. B. Freeman, 1975.

> Tell me someone why there's only confusion,
> Tell me someone that this is all an illusion,
> Tell me someone, tell me someone.
>
> Tell me someone why this talk of revolution,
> Tell me someone when we're changing evolution,
> Tell me someone, tell me someone.
>
> John Lodge, "One More Time to Live"

The chapter-opening quotation is from the Moody Blues record *Every Good Boy Deserves Favour*. © MCA Music Co., New York, 1971.

Chapter Twenty

GENES OF THE FUTURE

And so we arrive at the last chapter of the book. But are we also near the last chapter of our species? Paul Ehrlich, well known for urging people to defuse the population bomb, is by profession an evolutionary biologist. In 1970 he published a paper called "Looking backward from 2000 A.D.," a futuristic piece in which he pretended to be submitting a report from the federal Department of Population and Environment to the President of the United States as background for the President's state-of-the-union address in the year 2000. This report said the U.S. population was 22.6 million, but was no longer declining. Heroic efforts to decrease stillbirths and infant mortality through special reproducer rations and child-health squads seemed to be working. Two reasons cited for the small population were the release of enormous amounts of radioactivity from a nuclear reactor plant in 1981 and a great viral epidemic that resulted in about 125 million American deaths in the 1980's. The subsequent disorganization of medical care and of agriculture produced a famine which killed 65 million Americans in the decade 1980 to 1989. These calamities were compounded by previous atmospheric and water pollution from nuclear sources and industrial chemicals that acted as carcinogens and mutagens. By the 1990's, however, the government had relocated the surviving population away from the radioactive areas and virus-infested cities, and had turned the tide on the environmental deterioration that had been growing over the previous century. However, the government had not been able to counteract the bankrupt condition of ocean fisheries which resulted from contamination of the oceans with toxic pollutants washed down from the land.

The report recommended continued efforts to improve nutrition and environmental quality, and noted the need for more research in genetics, virology, and biochemistry in order to better understand, and hopefully control, the viral diseases and environmentally-induced mutational overload that almost fatally debilitated the American population.

EFFECTS OF MEDICAL TRENDS ON THE HUMAN GENE POOL

The quality of the environment of course has many effects on the gene pool, as the above scenario suggests. While the gene pool of future generations is surely at the mercy of nuclear catastrophes, chemical pollutants acting as mutagens on humans and viruses, and the level of basic health care provided by the medical profession, questions about the effects of some recent medical trends on future gene pools are also being raised. For instance, will genetic counseling affect the future pool? Although opinion is somewhat divided, it is generally thought that genetic counseling will have only a small impact on the pool because most genetic diseases are recessive and heterozygotes are so numerous (for all possible diseases) that persuading them not to be parents would preclude almost all reproduction.

What about the development of medical treatment to permit survival and reproduction of those whose medical defects formerly did not allow them to survive? Certainly here we have a situation of relaxed selection against hemophilia, sickle-cell anemia, diabetes and a host of other genetic or quasi-genetic disorders. Increasing treatment of hemophilia, for example, is now producing hemophiliac adults who are reproducing. Thus the frequency of hemophilia in the population will increase insidiously. But many geneticists have pointed out that the deterioration of the gene pool through these kinds of medical salvage operations is a very slow process, slower for the numerous recessive defects, of course, than for the rare dominant ones (Fig. 20–1).

Most people, including geneticists, find it difficult to sustain a feverish worry over these harmful spin-off effects of medicine, probably because the human mind is notoriously shortsighted. In this regard, James Watson (1971) has written, "Though scientists as a group form the most future-oriented of all professions, there are few of us who concentrate on events unlikely to become reality within the next decade

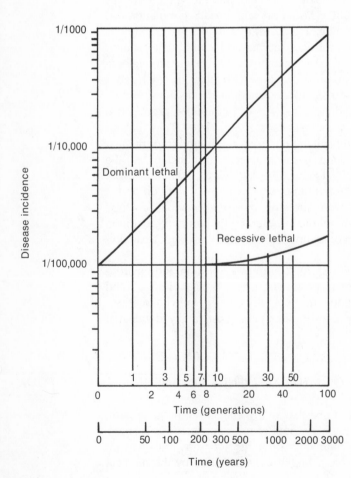

Figure 20–1 The effect of full relaxation of selection against previously lethal defects, assuming a pre-relaxation birth incidence of 1 in 100,000. (Modified from Bodmer, W., and Cavalli-Sforza, L.: Genetics, Evolution and Man. San Francisco, W. H. Freeman, 1976.)

or two." In the human past there has probably been strong selection for opportunistic genes, if interest-span or worry-span has any genetic basis at all. Most of human evolution has taken place under conditions of subsistence living, that is, in a state of mundane poverty. Oscar Lewis (1959), the anthropologist of Mexican life, has found that the culture of poverty shows "a strong present-time orientation and relatively little disposition to defer gratification and plan for the future."

If there has been a general lack of evolutionary pressure for "long thinking," our descendants may experience the day when half the population consists of doctors who take care of the other half, the patients with debilitating genetic defects. But that day is many generations away, and new technologies and crises may alter this forecast greatly. Remember also that for some of these debilitating conditions the heterozygote may have an advantageous phenotype in particular environments.

EFFECTS OF PHYSICAL COMFORTS

Related to the issue of gene salvage by medical treatment is the following question: do not the general comforts and safety of technological civilization produce a relaxation of natural selection and thus a deterioration of the human gene pool?

Certainly the physical environment was previously much more perilous than it is now. Given our present, more controlled environment, where famine, epidemics, and predators are rare, are we losing the ability to cope with physical stresses? Undoubtedly so to some degree. But only a small minority of the world's people enjoys lavish creature comforts, and it has done so only for a few generations. Moreover, most young people, at least, envision a somewhat harsher environment in the future—food and energy problems being what they are—so that many opportunities may again arise even in the United States to test the notion of physical survival under more stringent conditions. Also, as the British Nobel laureate Peter Medawar (1965) has pointed out, physical survival in the face of environmental stresses such as infectious disease, temperature extremes, and so on, does not denote the working of some virtuous gene for general survival. Each such survival gene is specific for one problem. The merit of the sickle-cell gene in combating malaria is a genetic trick that backfires in non-malarious environments (producing there only a grave anemia in the homozygote).

EFFECTS OF BREAKDOWN OF KINSHIP GROUPS

The close social bonds of kinship groups, still found in even the most undaunted liberation-prone societies, sometimes lead to inbreeding and, in traditional agricultural societies, are also associated with large families. Inbreeding brings out recessive defects which, unless they receive medical care, often cause death and thus loss of the defective genes, thereby minimizing the genetic load on the population. Large families bring about population growth, especially when medical care is available, and eventually lead to resource shortages and environmental doom. But large kindreds often provide a substantial amount of mutual aid. If there are genes for unselfish (altruistic) behavior, as has often been suspected, kin selection provides a way for such genes to be favored, since a kindred shares many genes in common. You may recall (Chap. Nineteen) that Garrett Hardin thought altruistic genes may now be selectively disadvantageous as a result of competition for the world's limited resources. In the context of the current shrinkage of close kin ties, resulting from smaller families and greater population mobility than formerly, any existing altruistic genes may be less able to be genetically effective. Thus we can envision two ways in which altruistic genes may be declining, through intensification of competition and through the shrinkage and dispersal of kin groups. It is more difficult to imagine selective pressures for their preservation.

EFFECTS OF TECHNOLOGY FOR SEX SELECTION

The Talmud says to align the marriage bed with the north-south axis in order to favor the begetting of sons. Presumably this method was not reliable, as the problem of sex preselection still exists today. A number of studies show that American adults want at least one boy in their family, preferably as the first child. (If sex preselection of offspring had been available to Henry VIII, he would have been able to assure himself of a male heir and probably would have led a less scandalous personal life.) Since in about half of all families the first child is a girl, would many people use some sort of technology—if it were available—to assure a first-born boy? The answer seems to be yes, if the technology were easy, cheap, and safe.

Although no such means of sex selection is at hand, it is already possible to determine the sex of one's offspring by the procedures of amniocentesis and karyotyping, and to perform selective abortion for an unwanted sex. However, medical-genetic centers, which perform such procedures, are unwilling to accede to requests for sex preselection, which they consider unwarranted and which involve procedures at least marginally more risky than an untampered pregnancy. A new development in 1975 was the discovery by Rhine and co-workers that cells shed by fetal tissues appear in the cervix early in pregnancy and can be stained selectively for the presence of the Y chromosome in the interphase state. This permits of an easy and early detection of fetal sex without amniocentesis. Other methods of selecting the sex of offspring may sometime be possible: sperm samples may ultimately be fractionated into X- and Y-types; antibodies against X- or Y-type sperm may be developed; a morning-after pill for sex preselection may be invented; and so on.

Another, more complex means of sex preselection is now possible, but it is still labeled experimental (Fig. 20–2). This procedure is for women who desire children but have defective Fallopian tubes in which a blockage prevents the fertilization of normal eggs and their subsequent travel down into the uterus. R. G. Edwards (1974) in England is developing the following treatment to overcome the sterility of these women: he and his associates administer a hormone to such a woman so that she sheds a number of eggs from her ovaries simultaneously. By surgical manipulation these eggs are collected and placed in laboratory containers suitable for their growth. The eggs are fertilized in vitro and their development monitored through the microscope. Those embryos with abnormal development are discarded. Studies up to this stage have been useful in providing information on conception and early human development, information hitherto not available. Edwards and others have further shown by animal studies that it is possible to reinsert a young embryo into the uterus of a "pregnant" mother, have it attach to the uterine wall, develop a placenta, and become a normal newborn after the usual period of pregnancy. Indeed, a trick used by animal breeders to "transport genes" is to reimplant fertilized eggs of prize-winning sheep inside a rabbit (as a living incubator), air-freight the rabbit from Britain to New Zealand, and have the eggs successfully reinserted into an ewe's uterus upon arrival.

Edwards has reinserted an in vitro human embryo into the uterus of its own mother, but without success in bringing it to term. But in 1974 there was a brief

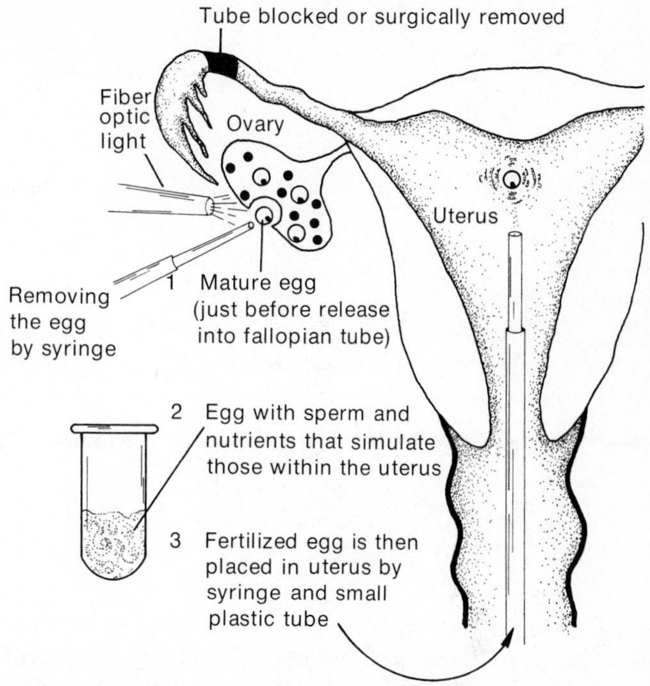

Figure 20–2 Achieving a detour for uterine implantation of a fertilized egg. (Modified from Time Magazine, July 29, 1974, p. 58.)

statement to the press by a British physician that such reinsertions had been done successfully on several occasions (see *Time Magazine* reference, "The Baby Maker"). No further announcements have been forthcoming. Not surprisingly, the physician was vigorously criticized by scientific and medical groups on various grounds. If the Edwards-kind of procedure became generally available to families, it could be used for sex preselection by removing a few cells from the early embryo for staining of the Barr (X) or Y body before the time of reinsertion.

Supposing that these and other means for sex selection may come to pass, let us consider the genetic and social implications for a boy-preferred world. The direct genetic fallout is minor: more boys would mean more appearances of X-linked recessive traits. The social fallout might be major: the sociologist Etzioni (1968), as well as others, pointed out that males are more competitive and more prone to criminal behavior than females. More Roman games to redirect aggression and more prisons would presumably be needed. An excess of males, Etzioni suggests, would produce a frontier stage set dominated by violence, prostitution, and homosexuality. But the law of supply and demand might also persuade parents to shift their offspring sex allegiance quickly, in an attempt to anticipate scarcity. Yet another consideration on sex preselection is that all sexes that eventuate would be "wanted," a boon for child acceptance and also for population limitations (since many parents want two children but have three or four in trying to get at least one of each sex). All in all, the easy availability of sex preselection would probably not much affect the human gene pool directly; indirect effects, however, are difficult to evaluate.

EFFECTS OF MOSAICISM AND CLONING

Among the many manipulative techniques not yet tried on human life are mosaicism and cloning. They have been accomplished on frogs, mice, and other test animals. Mosaicism refers to the presence of two or more genetic types of cells in a single individual (Chap. Six). Animals are usually mosaic to a minute extent owing

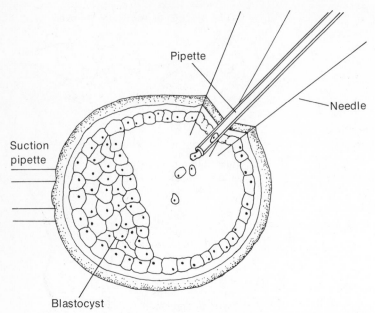

Pipette

Needle

Suction
pipette

Blastocyst

Figure 20–3 Colonizing an embryo with cells from another embryo, resulting in a combination of traits in the mosaic individual. (From Edwards, R.: Human embryos in the laboratory. Scientific American, 223:44–54, 1970. Copyright © 1970 by Scientific American.)

to spontaneous chromosome loss in a few cells, but mosaicism may also be contrived by the experimental emplacement of one or a few cells from one embryo into the embryonic cavity of another (Fig. 20–3). The invading cells grow and colonize various tissues, producing a genetic mosaic if the two embryos were genetically different.

One application of this technique is the infusion of colonizing cells from one person's bone marrow to the marrow area of a related infant who suffers from a genetic inability to produce bone-marrow derived lymphocytes and thus lacks immune responsiveness. Such children are then somatic mosaics. If this therapy allows such children to survive and reproduce, the technique can have an indirect effect on the gene pool if these mosaic individuals transmit their defective alleles to their offspring. By the same token, the successful transplantation of kidneys and other organs into genetically inviable individuals who survive and reproduce as a result of this therapy affects the gene pool. But this rarely occurs at present because of the formidable complications with the immune system in the survival of organ transplants.

Cloning refers to the production of identical copies of the same genotype. Identical twins comprise a two-person clone. Frog clones have been produced by removing the nuclei of fertilized eggs and replacing them with nuclei from more advanced embryonic cells of a different individual (Chap. Nine). The work of Edwards and others on the isolation of human eggs makes nuclear transplantation studies on human eggs a distinct possibility. James Watson predicted in 1971 that "if the matter proceeds in its current nondirected fashion, a human being born of clonal reproduction most likely will appear on the earth within the next twenty to fifty years, and even sooner, if some nation should actively promote the venture."

If cloning were ever tried on humans, and succeeded, a whole army or basketball team of genetically identical members could be produced. They might conceivably be very successful in their endeavors, being generally "of one mind." What the ultimate value of asexual reproduction by cloning would be for human existence is hard to say. Some commentators have suggested talented persons could someday have a "child" cloned from one of their body cells and spend the last years of their life teaching the child their specialty from their lifetime of experience. Presumably the child would be a "quick study."

The effect of cloning on the gene pool would be to narrow the genetic store of variability to a greater or lesser extent. From an evolutionary perspective, this would not be beneficial.

OTHER ASPECTS OF GENETIC ENGINEERING

Having now looked at the most likely avenues for human genetic engineering, such as sex preselection and cloning, I should mention that the technique of experimental hybridization of DNA from two sources inside a cell (see Box, Chap. Twelve, p. 139) could conceivably have an effect on the human gene pool. The technology for such DNA hybridization experiments is currently being reviewed for standards of safety in public health terms, but new developments are to be expected from such studies because of the large amount of research interest in the area.

Still other molecular means of genetic engineering exist as remote possibilities. One such means, the process of DNA transformation used in bacteria (Chap. Seven), may eventually be adapted for use on humans. For instance, Robert Sinsheimer has theorized that isolation of the normal gene for producing insulin (not yet possible) and its reinsertion into the chromosomes of pancreatic cells could provide a home-made source of the drug for diabetics.

AN ETHICAL PERSPECTIVE

Since, as H. G. Wells said, "we are not ourselves only, we are also part of human experience and thought," we have the perpetual responsibility to determine as best we can the effects of our actions and inactions. The professional responsibility of scientists is to search for factual truth about the world. They consider this to be an ethical activity. J. Bronowski said in this context, "The great ethical force of science has been proved to be the dissemination of the idea that truth is a thing which will in some way help us all." The way truth helps us all is primarily by informing us of the complex reality of the world. Indeed, all of education functions to make us aware of complexity and "mind-expanding" reality.

Because nature—including human nature—is complex, simple answers to problems about nature are usually not fully truthful. One might think that the answer to problems of pollution, hungry nations, the nuclear war threat, medical overtreatment, destruction of wilderness, and other favorite topics of concerned biologists is to revert to a "natural" condition in which humanity interferes as little as possible with the great design of nature. Indeed, retreat from technological peril is sometimes a tenet of today's newly rural philosopher-farmers. But, of course, only farming which uses no motors or machinery, which maintains soil fertility by crop rotation and manuring, and which prevents soil erosion permanently is ecologically sound. Moreover, there is not enough arable land for everyone to become a subsistence farmer, even if they could learn how. And the breakdown of social ties that such ruralized dispersion would bring about presents an insuperable obstacle for most people. Other difficulties with this idea easily come to mind. But as a general rule, treading lightly on the world is, I think, rational and necessary.

We cannot deny that today humanity does stand at the pivotal point of nature. This position is not, of course, the same as the medieval belief in man's having been put on earth, the center of the universe, by God. But the human species today is no longer a marginal one eking out a chancy day-to-day existence, at least not in

Western countries. We have profound power over nature, the power to destroy it and ourselves unwittingly. Why does the world allow this human danger to arise? How we as a species escaped, at least in the short run, the "checks and balances" seen controlling other populations is not fully known. While our large brains are involved in our potential demise, from a bioethical view they are also our hope of salvation.

Should Western nations accept the so-called technological imperative of saving the world? I. M. Lerner (1975), the evolutionary geneticist, says no: "At first blush, according to any reasonable moral standard, we, who have so much, must help our less privileged fellow men to achieve a longer, more comfortable, happier life. . . . To do nothing but watch hunger and pestilence rampant obviously does not fit [the humanitarian] ethic. But what are we really doing? We are helping the developing countries to increase their populations, to accelerate depletion of their resources, to strain their carrying capacity to an intolerable level, if not in the current generation" (p. 26). But, you say, does not a Yale or Stanford graduate consume more resources (including guns for war and plastic for phonograph records) than any one person in hungry Asia? You are correct. But the fact of the matter is that if, say, Americans lived at the material level of an average Asian and devoted their "extras" to poor people, the giveaways would only produce a quick surge in population growth, given the poor state of birth control technology and the social hindrances to its usefulness today. The next generation would be more numerous and thus individually even worse off.

Should we then do nothing? Biologists tend to agree that both action and inaction concerning the world's problems seem to produce unwanted effects, even in the crucial area of birth control. We do not know enough to evaluate different possibilities in depth. This is the central problem of human existence now. There seems to be no ready solution.

FUTURE HUMAN EVOLUTION

The long future of the human species depends on many factors mentioned in this text. The main point to remember is that, barring extinction, evolutionary changes are slow ones. Skeletons of persons buried 10,000 or more years ago are indistinguishable from the human skeleton today. Skeletons 10,000 years hence may also be indistinguishable from those of today. Indeed, most aspects of the human form change too slowly to allow us to measure the rate of change in one century. It may take a thousand years of record-keeping to detect significant general changes in human anatomy or physiology. Record-keeping on the genetic effects of environmental mutagens, on disease control, and the rampant breakdown of cultural isolation that fosters gene flow among groups may, however, show significant changes in the human gene pool on a time scale shorter than 1000 years — perhaps even by the twenty-second century. This current and future gene flow has the potentiality for evolving new gene complexes adaptive for future environments, regardless of how rich or austere they may be.

SUMMARY

This gloomy chapter has considered many dire consequences of our technological way of life. Although we cannot dictate what the future will be, we possess the attributes of worry, hope, and rational thought — which in themselves permit of some slight optimism.

REFERENCES*

Edwards, R. G. Fertilization of human eggs in vitro: morals, ethics, and the law. Quarterly Review of Biology *49*:3–26, 1974.

Ehrlich, P. Looking backward from 2000 A.D. The Progressive, April, 1970, pp. 23–25.

Etzioni, A. Sex control, science, and society. Science *161*:1107–1112, 1968.

Lerner, I. M. Ethics and the new biology. *In* Genetics and the Quality of Life, C. Birch and P. Abrecht (eds.). New York: Pergamon Press, Inc., 1975, pp. 20–35.

Lewis, O. Five Families: Mexican Case Studies in the Culture of Poverty. New York: Basic Books, Inc., 1959.

Medawar, P. B. Do advances in medicine lead to genetic deterioration? Mayo Clinic Proceedings *40*:23–33, 1965.

Rhine, S. A., J. L. Cain, R. E. Cleary, C. G. Palmer, and J. F. Thompson. Prenatal sex detection with endocervical smears: successful results utilizing Y-body fluorescence. American Journal of Obstetrics and Gynecology *122*:155–160, 1975.

Sinsheimer, R. L. The prospect of designed genetic change. Engineering and Science, April, 1969, pp. 8–13.

The baby maker. Time Magazine, July 29, 1974, pp. 58–59.

Watson, J. D. Moving toward the clonal man. Atlantic Monthly *227*:50–53, 1971.

SUGGESTED READINGS

Ausabel, F., J. Beckwith, and K. Janssen. The politics of genetic engineering: Who decides who's defective? Psychology Today, June, 1974, pp. 30–43.

Austin, C. R. Embryo transfer and sensitivity to teratogenesis. Nature *244*:333–334, 1973.
Austin discusses the hazards of in vitro fertilization for human existence.

Bennett, D., and E. Boyse. Sex ratio in progeny of mice inseminated with serum treated with H–Y antiserum. Nature *246*:308–309, 1973.
Sex ratio can be somewhat controlled in mice by antibody treatment of sperm.

Berns, M. W. Directed chromosome loss by laser microirradiation. Science *186*:700–705, 1974.
Lasers may permit chromosome engineering.

Edwards, R. G. Human embryos in the laboratory. Scientific American *223*:44–54, 1970.

Edwards, R. G., and D. J. Sharpe. Social values and research in human embryology. Nature *231*:87–91, 1971.

Ericsson, R. J., C. N. Langevin, and M. Nishino. Isolation of fractions rich in human Y sperm. Nature *246*:421–424, 1973.

Friedmann, T., and R. Roblin. Gene therapy for human genetic disease? Science *175*:949–955, 1972.

Fuhrmann, W. Therapy for genetic diseases in man and the possible place of genetic engineering. Advances in BioScience *8*:387–395, 1972.

Haldane, J. B. S. Biological possibilities for the human species in the next ten thousand years. *In* Man and His Future, G. Wolstenholme (ed.). Boston: Little, Brown & Co., 1963.

Jones, A., and W. F. Bodmer. Our Future Inheritance: Choice or Chance? London: Oxford University Press, 1974.
A nontechnical report from a British group.

Kass, L. R. Babies by means of in vitro fertilization: unethical experiments on the unborn? New England Journal of Medicine *285*:1174–1179, 1971.

Kaufman, M. H., E. Huberman, and L. Sachs. Genetic control of haploid parthenogenetic development in mammalian embryos. Nature *254*:694–695, 1975.
Haploid hamster embryos have developed in vitro for several days.

Lockard, J. S., L. L. McDonald, D. A. Clifford, and R. Martinez. Panhandling: sharing of resources. Science *191*:406–408, 1976.

Marx, J. L. Embryology: out of the womb—into the test tube. Science *182*:811–814, 1973.

Murdy, W. H. Anthropocentrism: a modern version. Science *187*:1168–1172, 1975.
Our survival depends on a belief in human creativity.

Steptoe, P. C., and R. G. Edwards. Reimplantation of a human embryo with subsequent tubal pregnancy. Lancet *1*:880–882, 1976.

Westoff, C. F., and R. R. Rindfuss. Sex preselection in the United States. Science *184*:633–636, 1974.

Wilson, E. O. Sociobiology: The New Synthesis. Cambridge, Mass.: Harvard University Press, 1975.

---

*Note: items cited in the References are also recommended for suggested reading.

Appendix A

MORE MENDELIAN GENETICS

The allele segregation ratios of 3:1 and 1:1 in simple crosses that Mendel worked out are often referred to as instances of "Mendel's first law." This generalization states that the particles of heredity for specific characteristics separate cleanly during gamete formation and reunite at random in zygote formation (as discussed in Chapter Two).

INCOMPLETE DOMINANCE

Mendel also observed other genetic phenomena, one of which is the condition of incomplete dominance. Incomplete dominance refers to instances when the phenotype of the heterozygote falls between the phenotypes of the two corresponding homozygotes. Mendel studied the cross of red flowered *Mirabilis* with white flowered *Mirabilis* (commonly called four-o'clocks). With the symbol A_1 for the red allele and A_2 for the white, the cross can be diagrammed as

P: A_1A_1 × A_2A_2

(red)   (white)

F_1: A_1A_2

(pink)

Here we use only capital letters for allele symbols because neither A_1 nor A_2 is dominant. For *Mirabilis* the F_2 produced by crossing together two pink F_1's has a phenotypic ratio of 1:2:1 instead of the 3:1 ratio exhibited in cases of full dominance.

Many cases of incomplete dominance are known for a variety of traits in diverse species, including humans. Extensive studies have shown that the heterozygote in incompletely dominant situations need not be *exactly* intermediate between the two homozygous phenotypes. Indeed, examples are known that differ from one or the other homozygote only to a slight extent. A further expansion of the idea of dominance is presented in the following section.

OVERDOMINANCE: HYBRID VIGOR

The term overdominance refers to simple heterozygote phenotypes that are neither intermediate between the homozygous types (incomplete dominance) nor exactly the same as one of the homozygotes (dominance), but in fact exceed the phenotype of the "dominant" homozygote. As this statement suggests, these various terms about dominance are not discrete alternatives but form a continuum from more to less dominance—or from more to less recessiveness, to put it the other way around (Fig. A–1).

If P genotypes: AA × aa

Then F₁ genotype: Aa

But F₁ phenotype depends on degree of dominance expressed for each   particular gene trait:

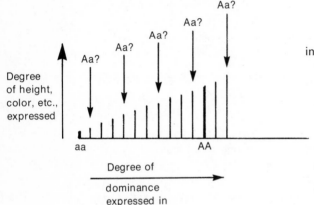

Figure A–1   Various degrees of dominance are possible in heterozygotes (Aa).

Mendel described a case of overdominance. One of the traits he studied in garden peas was the height of the full grown plant. The true-breeding strains tall and dwarf, which had heights of 6 feet and 1 foot, respectively, when crossed together gave F₁ offspring varying in height from 6 to 7.5 feet. This overdominance of the heterozygote is often called hybrid vigor. Although it has not been definitively explained in terms of how genes determine phenotypes, hybrid vigor is worthy of attention because it is observed often in plant and animal breeding studies for agricultural improvement (see Chap. Eighteen).

INDEPENDENT ALLELE SEGREGATION FOR SEPARATE GENES

The so-called second law of Mendel explains the phenotypic ratios obtained when true-breeding strains differing in two or more independent characteristics are crossed together. After studying several two-trait crosses, Mendel concluded that the alleles for one trait segregate independently of the alleles for the other trait under observation because he obtained phenotypic ratios that were expected on this basis. For example, in the pea cross of a tall, yellow-seeded double heterozygote with a true-breeding dwarf, green-seeded plant, offspring are obtained that are approximately 25% tall and yellow, 25% tall and green, 25% dwarf and yellow, and 25% dwarf and green. This is a 1:1:1:1 phenotypic ratio:

genotypes	Tt Yy × tt yy	
phenotypes	(tall, yellow)	(dwarf, green)
gametes	$\frac{1}{4}$ TY	all ty
	$\frac{1}{4}$ Ty	
	$\frac{1}{4}$ tY	
	$\frac{1}{4}$ ty	

offspring

$\frac{1}{4}$ Tt Yy (tall, yellow)

$\frac{1}{4}$ Tt yy (tall, green)

$\frac{1}{4}$ tt Yy (dwarf, yellow)

$\frac{1}{4}$ tt yy (dwarf, green)

Such experimental results fit Mendel's idea that the segregation to the gametes of alleles for one trait (say, T vs. t) is an event independent of the segregation of the alleles for another trait (here, Y vs. y). That is, T is just as likely to occur in a gamete with Y as it is in one with y—there is no partiality in either direction. This result occurs because the genes Mendel observed were located on different chromosomes (see Chap. Five).

Perhaps a more famous demonstration of the independent segregation idea is the appearance of a 9:3:3:1 ratio among the offspring of a cross between two double heterozygotes. To continue with the pea traits above, consider the following cross:

genotypes	Tt Yy × Tt Yy	
phenotypes	(tall, yellow)	(tall, yellow)
gametes	$\frac{1}{4}$ TY	$\frac{1}{4}$ TY
	$\frac{1}{4}$ Ty	$\frac{1}{4}$ Ty

$$\frac{1}{4} \, tY \qquad\qquad \frac{1}{4} \, tY$$

$$\frac{1}{4} \, ty \qquad\qquad \frac{1}{4} \, ty$$

These four kinds of ovules and pollen can be shown to generate 16 offspring compartments in a Punnett square; the results simplify to four phenotypic classes in the ratio 9:3:3:1.

	TY	Ty	tY	ty
TY	TTYY	TTYy	TtYY	TtYy
Ty	TTYy	TTyy	TtYy	Ttyy
tY	TtYY	TtYy	ttYY	ttYy
ty	TtYy	Ttyy	ttYy	ttyy

■ = 9/16 tall yellow

□ = 3/16 tall green

□ = 3/16 dwarf yellow

◻ = 1/16 dwarf green

The 9:3:3:1 ratio thus demonstrates the independent segregation and random recombination of alleles of different genes in forming the next generation, just as the 1:1:1:1 ratio mentioned earlier does.

A ratio different from 9:3:3:1 is obtained in such crosses of double heterozygotes if, instead of complete dominance for each trait, one or both of the genes segregate alleles that are incompletely dominant, or if the heterozygote is overdominant. Such modifications of the 9:3:3:1 ratio in a two-gene cross can produce surprising numerical results (e.g., 1:2:1:2:4:2:1:2:1 for the case of two genes where each shows incomplete dominance).

INTERACTION EFFECTS OF DIFFERENT GENES

Although genes may be non-interactive, or independent, in their segregation patterns to gametes, they may also be noticeably interactive in the effects they have on phenotypes. An often described example of such phenotypic interaction is the situation in which one gene has an inhibitory effect on the product of the other gene. Consider the situation where allele A (dominant) specifies color in certain flowers, and its allele a (recessive) specifies colorless (white). All other things being equal, we would then expect that the cross Aa × Aa (color × color) would produce $\frac{3}{4}$ colored and $\frac{1}{4}$ white offspring. But imagine that this plant also has an independently segregating gene where allele B specifies an inhibitory effect on color but b has no inhibitory effect (is recessive to B). The full genotypes of the cross color × color mentioned above would then be Aa bb × Aa bb; but since bb does not show allele segregation here, we would still get a 3:1 phenotypic ratio in the offspring.

If, however, the genotypes crossed together were Aa Bb × Aa Bb (both parents inhibited by B from expressing color), the expected offspring ratio is then:

9 A– B–, white
3 A– bb, color
3 aa B–, white
1 aa bb, white

or a ratio of 13 white to 3 color. Ratios such as 15:1, 12:3:1, and other derivatives of 9:3:3:1 have been shown to occur in particular cases of interaction of gene effects among a wide sampling of crosses in many organisms.

SUMMARY

This discussion has extended the basic findings of Mendel in several directions. These extensions include conditions of incomplete and overdominance, the independent segregation of alleles of different genes, and the interaction effects of independently segregating genes. Other extensions of Mendelian genetics are discussed elsewhere in the book.

STUDY QUESTIONS

1. Using a Punnett square, verify the 1:2:1 ratio produced from the pink × pink *Mirabilis* cross discussed in the text.
2. What is the expected phenotypic ratio in the progeny of the cross Tt yy (tall, green) × tt Yy (dwarf, yellow)? What is the reason that not all the offspring are yellow? That not all are tall? That not all are both yellow and tall?
3. What is the expected phenotypic ratio in the progeny of the cross Tt yy (tall, green) × tt yy (dwarf, green)?
4. Key terms:
   incomplete dominance   hybrid vigor
   overdominance   independent allele segregation

Appendix B

GLOSSARY

ABO blood group
This blood group system, polymorphic in human populations, is determined by a gene which has multiple alleles. The various allelic combinations specify an antigenic configuration on the surface of red blood cells as well as certain antibodies in the blood serum. (Chap. 11.)

Achondroplasia
A form of dwarfism; an autosomal dominant trait. (Chaps. 13, 15.)

Acidic proteins
A mixture of proteins associated with chromosomes that have an acid character. Contrast: *histones*. (Chap. 9.)

Albinism
The phenotype of a deficiency or near deficiency in melanin pigmentation in the skin, the eye, and other tissues; a recessive autosomal trait. (Chap. 3.)

Alcaptonuria
A recessive autosomal trait characterized by urine which darkens upon exposure to air. (Chap. 10.)

Allele
A particular (chemical) state of a gene (DNA segment). A gene may have one or more alternative alleles, each allele determining a particular form of the gene product, or phenotype. New alleles may arise through mutation. (Chap. 2.)

Allele segregation
See *segregation*.

Amino acid
Any of the 20 or so small molecules characterized by both acidic and basic properties, resulting from the chemical components —COOH and NH_2—, respectively. Amino acids are the component parts of proteins. (Chap. 7.)

Amniocentesis
The process by which fluid is drawn from the amniotic sac surrounding the fetus in the uterus. This amniotic fluid is then tested by a variety of methods for the presence of chemical or chromosomal abnormalities. (Chap. 6.)

Anaphase
The stage in mitosis and meiosis which follows metaphase and precedes telophase. During anaphase of meiosis I homologous chromosomes separate to opposite poles of the cell. During mitotic anaphase and the second meiotic anaphase the two chromatids of each chromosome separate to opposite poles. (Chap. 4.)

Aneuploidy
Any chromosomal constitution which deviates in number of chromosomes from a full haploid or diploid set (e.g., trisomy). (Chap. 6.)

Antibody
One of the set of proteins circulating in the bloodstream that has the capacity to attach to and thereby inactivate a foreign antigen, thus producing an immune response. Each kind of antibody is specific for one kind of antigen. (Chap. 11.)

Anticodon
The three-base sequence on a transfer RNA molecule that base pairs with the codon sequence on messenger RNA. Such pairing is required to bring about the insertion of a specific amino acid into a chain of amino acids forming a protein molecule. (Chap. 8.)

Antigen
One of the types of chemical substances to which a specific antibody can attach, thereby producing an immune response. Each kind of antigen determines, in some unknown way, the production of a single kind of antibody which can react with that antigen. (Chap. 11.)

Atom
A component of a molecule. Atoms are of many different kinds, e. g., carbon (C), hydrogen (H), oxygen (O). Combinations of similar or different atoms form molecules. For example, two atoms of oxygen (O) form a molecule of oxygen (O_2). (Chap. 7.)

Autoimmunity
The formation of antibodies to the body's own antigens, i.e., to oneself. Certain diseases are thought to be caused in this way. (Chap. 11.)

Autosome

Any chromosome other than the sex chromosomes (X, Y). (Chap. 6.)

Balanced polymorphism

A polymorphism in which alternative alleles are maintained at intermediate frequencies in the population by the heterozygote being selectively advantaged in comparison to the homozygotes, or by other means. (Chap. 17.)

Barr body

The dark-staining X-chromosome mass seen in interphase nuclei. Generally, the number of Barr bodies per nucleus is one fewer than the number of X chromosomes in the nucleus. Consult: *X-inactivation*. (Chap. 6.)

Base analogs

Chemical compounds similar in shape and hydrogen-bonding properties to the bases normally found in DNA and RNA. (Chap. 13.)

Base pair

The particular purine-pyrimidine combinations held together by hydrogen bonds between strands of nucleic acids. The usual base pairs are: AT, TA, GC, and CG in DNA; and AU, UA, GC, and CG in RNA. (Chap. 7.)

Base pair substitution

The replacement of one base pair by another in DNA, usually as a result of mutation. (Chap. 13.)

Bases (in DNA and RNA)

The nitrogen-containing components of DNA and RNA, normally adenine (A), guanine (G), cytosine (C), and thymine (T) in DNA, and A,G,C, and uracil (U) in RNA. (Chap. 7.)

Biochemical pathway

A series of enzymatic reactions in life forms that transforms one kind of molecule by small steps into another kind. (Chap. 10.)

Blood type

The immunological phenotype of an individual expressed by the antigens on his/her red blood cells and the antibodies to these red-cell antigens found in the blood serum. (Chap. 11.)

Bottleneck

A severe contraction in the size of the gene pool, to the point where genetic variability is substantially reduced in the population. (Chap. 16.)

Cancer

Any of a number of diseases in which there is uncontrolled cellular growth. (Chap. 12.)

Carbohydrates

Starches and sugars. (Chap. 7.)

Carcinogen

Any agent that induces cancer, e.g., X-rays, certain chemicals. (Chap. 12.)

Carrier

One who has an allele but does not express its effects.

Cell

The elementary unit of an organism, containing nucleus and cytoplasm and bounded by a membrane. (Chap. 3.)

Cell transformation

The process of cell change in which the cell loses its ability to control its rate of division and thus becomes a tumor cell. (Chap. 12.)

Centrioles

The two small organelles lying outside the nucleus that migrate to opposite poles of the cell in prophase; they are the termination points of the spindle fibers. (Chap. 4.)

Centromere

The constricted region of a chromosome to which the spindle fiber is attached during mitosis and meiosis. (Chap. 4.)

Chemical bond

The various attractive forces that exist between atoms or ions in molecules. E.g., see: *hydrogen bond, covalent bond*. (Chap. 7.)

Chiasma

The X-shaped region of a chromosome pair seen in prophase of meiosis I. Chiasmata are the result of interchanges (cross overs) between homologous chromosomes. (Chap. 5.)

Chromatid

The major longitudinal subunit of a chromosome. The two chromatids of a chromosome are held together at the centromere. After the centromere divides, each chromatid is a separate chromosome. Following a further round of synthesis of chromosomal material, the chromosome again consists of two chromatids. (Chap. 5.)

Chromosomal segregation

The process of separation of homologous chromosomes, and thus their constituent alleles, to different gametes during meiosis. (Chap. 5.)

Chromosome

A linear structure, consisting of DNA and proteins, which contains a sequence of genes. In nondividing cells the chromosomes are indistinct in the nucleus. Chromosomes become progressively condensed, and thus distinguishable, as a cell enters mitosis or meiosis. (Chap. 3.)

Clone

A group of cells all derived mitotically from one cell and thus genetically identical. (Chaps. 11, 14.)

Codominance

The condition in which, in a heterozygote, both alleles are expressed independently. (Chap. 11.)

Codon

A sequence of three bases, in DNA or RNA, which specifies a particular amino acid in protein or which specifies termination or initiation of transcription. (Chap. 8.)

Color blindness

Inability to see one or more colors; the usual form of color blindness is an X-linked recessive trait with defects in the perception of red and green colors. (Chap. 5.)

Concordance
The condition when both twins in a set are found to have a certain trait. Contrast: *discordance*. (Chap. 14.)

Conditional lethal
A genetic state in which the phenotype produces death under some conditions but not others. (Chap. 6.)

Consanguinity
Of the same "blood," i.e., genetic relatives. (Chap. 14.)

Covalent bond
The attractive force between atoms or ions in molecules characterized by the sharing of electrons. (Chap. 7.)

Cross-fertilization
The condition in which male gametes from one individual fertilize female gametes of another individual; self-fertilization is the contrasting condition, particularly among plants. (Chap. 2.)

Crossing over
The process of interchange of segments between homologous chromosomes. Cross overs produce new genetic combinations along the length of chromosomes. (Chap. 5.)

Cystic fibrosis
An autosomal recessive disease prevalent in the white population that is characterized by the accumulation of mucus in various organs. (Chap. 10.)

Cystinuria
An inborn error of metabolism characterized by painful deposits of material in urinary tissues. (Chap. 10.)

Cytoplasm
The part of the cell between the cell membrane and the nucleus; it contains ribosomes, mitochondria, and other organelles of importance for cell maintenance and growth. (Chap. 4.)

Darwinism
The notion, credited to Charles Darwin, of change in species over time due to natural selection; i.e., implies the greater success of certain hereditary variants in adapting to the environment. (Chap. 17.)

Deletion
Loss of a portion of a chromosome or, at the molecular level, of one or more base pairs in a DNA sequence. (Chap. 6.)

Diploid
The condition of a cell having a chromosome number twice that of gametes, which have a haploid chromosome number. (Chap. 5.)

Discordance
The condition in which one twin has a particular phenotype and the other one does not. (Chap. 14.)

Dizygotic
Referring to fraternal twins, that is, those derived from separate fertilized eggs. Contrast: *monozygotic*. (Chap. 14.)

DNA (deoxyribonucleic acid)
The genetic material of all cells and some types of viruses. (Chaps. 3, 7.)

Dominance
The condition in which, in a heterozygote, the action of one allele masks the phenotypic effect of the other, thus making the heterozygote phenotype the same as that of the relevant homozygote. The masking allele is said to be the cominant one. Contrast: *codominance, incomplete dominance, overdominance, recessiveness*. (Chap. 2.)

Double helix
The architectural structure of DNA. A DNA double helix is two "spiral" staircases in tandem. (Chap. 7.)

Down's syndrome
A clinical condition caused by an excess of all or part of a number 21 chromosome. It is characterized by mental retardation and several physical abnormalities. (Chap. 6.)

Drift (genetic drift)
The process of random changes in allele frequencies brought about by chance deviation in the distribution of alleles in meiotic segregation and in fertilization, or by other chance events. (Chap. 16.)

Duplication
Gain of a portion of a chromosome or, at the molecular level, of one or more base pairs in a DNA sequence. (Chap. 6.)

Electrophoresis
The method of separating large molecules, particularly proteins, on the basis of their rate of migration through a supporting medium in an electrical field. (Chap. 10.)

Enzyme
A protein which is capable of catalyzing a specific chemical reaction. Different enzymes catalyze different reactions in cells. (Chap. 7.)

Eukaryote
Any organism in which the cell nucleus has a nuclear membrane, i.e., most plants and animals. Contrast: *prokaryote*. (Chap. 9.)

Evolution
The notion of evolution, or change, of life forms over time is a very old one. Natural selection, as the most important mechanism for this change, was first publicized by Charles Darwin. (Chaps. 10, 16.)

Fertilization
Union of gametes of opposite sex to produce a zygote. (Chap. 2.)

Fitness
The relative success of an individual in transmitting genes to the next generation. (Chap. 17.)

Founder effect
The situation in which a small group of related

individuals founds an isolated population. In consequence of the group's size, it contains only a small, nonrepresentative sample of the genes in the population emigrated from. Thus, the descendants of the small group will also contain a distinctive sample of genes. (Chap. 16.)

Frameshift
Pertains to the frame of reading of the codon sequence during protein synthesis. If a base pair has been added or deleted in the DNA base sequence, this produces a frameshift mutation at the protein level. (Chap. 13.)

Gamete
A mature sex cell, sperm or egg. As a rule, gametes have a reduced (haploid) number of chromosomes. (Chaps. 2, 5.)

Gene
A segment of DNA (RNA in some viruses) that codes for a particular molecular product: a protein molecule or an RNA molecule. (Chap. 2.)

Gene pool
All the genes possessed by individuals in a population, considered together. (Chap. 16.)

Genetic code (for proteins)
The correspondence between the codons in DNA or RNA and the particular amino acids in the protein product. (Chap. 8.)

Genetic drift
See *drift.*

Genetic engineering
The term generally applied to any process of human modification of genetic systems at the molecular level. (Chap. 18.)

Genetic heterogeneity
The condition resulting from the production of the same phenotype, or very similar phenotypes, by more than one kind of genotype within a population. (Chaps. 10, 14.)

Genetic load
The total of disability and death in a population caused by genetic factors. (Chap. 17.)

Genotype
The genetic nature of an individual with reference either to a particular gene or to the complete gene complement. Compare: *phenotype.* (Chap. 2.)

Germ cells
Cells which are part of the reproductive tissue and which ultimately become sperm or egg cells. (Chap. 5.)

Germinal mutation
A mutation which occurs in the germ cells. Contrast: *somatic mutation.* (Chap. 13.)

G6PD (glucose-6-phosphate dehydrogenase)
An enzyme found in red blood cells and other tissues that catalyzes one step in the biochemical breakdown of sugar. Genetic variants of G6PD are known, some of which cause a form of anemia. (Chap. 9.)

Green revolution
A general term used to refer to any recent advances in plant breeding—particularly hybridization of crop plants—that are associated with increased yields of field crops. (Chap. 18.)

Haploid
The condition of a cell having a chromosome number characteristic of gametes, i.e., half the diploid number. (Chap. 6.)

Hardy-Weinberg model of nonevolution
The statement that the genotype frequencies in a large, randomly mating population will remain constant from generation to generation unless mutation, natural selection, migration, or genetic drift intervenes. (Chap. 16.)

Hemoglobin
The protein in red blood cells which transports oxygen from the lungs to the body tissues. It consists of an iron component, to which oxygen attaches, and four protein chains. In hemoglobin A, the commonest kind in the human body after birth, there are two alpha and two beta chains. In fetal hemoglobin there are two alpha and two gamma chains. In addition to such developmental differences, a variety of genetic variants occur within human populations. (Chaps. 8, 10.)

Hemophilia
A genetic defect in blood clotting ability; an X-linked recessive trait. (Chap. 5.)

Heritability
The estimate of the relative influence of genes and environmental factors in determining the average phenotype in the population, under a given set of environmental circumstances. (Chap. 14.)

Hermaphrodite
A person with both male and female reproductive tissue. (Chap. 6.)

Heterozygotic
(1) Having different alleles for a particular gene. (2) Having different structural arrangements between homologous chromosomes. Contrast: *homozygotic.* (Chaps. 2, 6.)

Histones
A class of chromosomal proteins which are rich in basic amino acids such as lysine or arginine. Contrast: *acidic proteins.* (Chap. 9.)

Hominid
In the human line of evolution. (Chap. 19.)

Homologous chromosomes
Chromosomes alike, or very similar, in structure and genetic functions. Such chromosomes pair in meiosis and contain genes governing the same traits. (Chap. 5.)

Homozygotic
Having the same alleles for a particular gene. Contrast: *heterozygotic.* (Chap. 2.)

Hormone
Any of the class of molecules which is synthe-

sized in one tissue of the body but stimulates cellular activity in a different tissue or tissues. (Chap. 9.)

Hybrid
The offspring resulting from crossing together two strains which are phenotypically (and genetically) different. (Chap. 2.)

Hybrid vigor
The condition in which, in a hybrid, there is greater luxuriance of growth or other genetic traits than in either parental line. Hybrid vigor presumably results from extensive heterozygosity. (Chap. 18.)

Hydrogen bond
The attractive force between atoms or ions characterized by the sharing of a hydrogen atom. (Chap. 7.)

Hyperplasia
Excessive cell proliferation. (Chap. 12.)

Immune response
The interaction of antigen with its specific antibody. (Chap. 11.)

Inborn error
A genetic defect in biochemical functioning evident at birth; usually the result of an enzyme deficiency for a particular step in metabolism. (Chap. 10.)

Inbreeding
Marriage between relatives, i.e., between individuals who possess more genes in common than two individuals chosen at random. (Chaps. 3, 16.)

Incompatibility
Lack of mutual tolerance on the biochemical level, particularly with respect to immunological traits. (Chap. 11.)

Incomplete dominance
The condition in which, in a heterozygote, the effect of one allele partially masks the phenotypic effect of the other allele. (App. A.)

Independent segregation (independent assortment)
The condition in which one chromosome pair segregates homologs to opposite poles of the cell without regard to the manner of segregation of other chromosome pairs. Thereby, genes on different chromosome pairs also segregate independently of each other. (Chap. 5.)

Inducer
A small molecule which participates in activating a gene to transcribe an RNA copy. (Chap. 9.)

Interphase
The portion of the cell division cycle between telophase and prophase during which DNA and other cell components are synthesized. (Chap. 4.)

Inversion
The reverse arrangement of a portion of a chromosome or, at the molecular level, of one or more base pairs in a DNA sequence. (Chap. 6.)

In vitro
Experimentation on biological systems outside the intact organism or cell.

In vivo
Experimentation on intact biological systems.

Isozymes
Distinguishable variants of the same enzyme in different tissues of an individual. (Chap. 10.)

Karyotyping
The process of preparing a photographic reproduction of a metaphase chromosome array and arranging the photographed chromosomes linearly by size. (Chap. 6.)

Klinefelter's syndrome
The medical designation for the set of characteristics found in individuals who possess two X chromosomes and one Y chromosome. (Chap. 6.)

Lactose intolerance
The inability to metabolize milk sugar (lactose). (Chap. 19.)

Lamarckianism
The notion, associated with the name of Jean Lamarck, that species change over time as a result of the inheritance of characteristics acquired by nongenetic adjustment to environmental conditions. (Chap. 17.)

LDH
Acronym for lactic acid dehydrogenase, the well-studied enzyme which is known to differ in different tissues of individuals. (Chap. 10.)

Lesch-Nyhan disease
A rare enzyme-deficiency disease which is lethal at any early age as a result of metabolic and behavioral abnormalities. (Chap. 14.)

Lethal genes
Those genes which are incompatible with survival at some stage prior to reproduction. Recessive lethals require two doses of the gene in question for lethality to be produced; dominant lethals require only a single dose of such a gene. Compare: *conditional lethal*. (Chap. 6.)

Linkage
The condition in which two or more genes are physically associated on the same chromosome. Linked genes tend to segregate together in meiosis. (Chap. 5.)

Lipids
Fat-like chemical substances. (Chap. 7.)

Lysenkoism
The anti-Mendelian doctrine in Russia which caused great harm to biologists and their science. (Chap. 17.)

Lysis
The breaking open of a cell as caused by a virus, for example, or by other agents. (Chaps. 11, 12.)

Lysogeny
A property of some viruses. Lysogenic viruses do not, with rare exceptions, lyse cells; rather, their chromosome becomes inserted into a

chromosome in the host cell nucleus. (Chap. 12.)

Malaria

An infectious disease caused by a single-celled animal, *Plasmodium,* which is transmitted from person to person by the bite of certain kinds of mosquitos. The disease is characterized by the periodic fever and chills which result from periodic invasion of red blood cells by the parasitic *Plasmodium,* the subsequent multiplication of the parasite in these cells, and the lysis of the cells by the parasite. (Chaps. 10, 17.)

Meiosis

The specialized form of mitosis that occurs during two cell divisions in the development of mature sex cells. By the process of meiosis the chromosome number per cell is reduced from diploid to haploid. Compare: *mitosis.* (Chap. 5.)

Mendelian

Pertaining to traits which show segregation patterns in conformity with the principles worked out by Gregor Mendel. (Chap. 2.)

Messenger RNA (mRNA)

The transcript of the DNA of those genes which code for proteins. (Chap. 8.)

Metabolism

The sum of the enzyme-mediated, biochemical steps which bring about the breakdown of food substances, the synthesis of cellular components, and the production of waste materials. (Chaps. 7, 10.)

Metaphase

The stage in mitosis and meiosis which follows prophase and precedes anaphase. During metaphase of meiosis I chromosome pairs are located in the midplane of the cell; during metaphase of mitosis and of meiosis II single chromosomes are located in the midplane. (Chap. 4.)

Metastasis

The spread of cancer cells from their site of production to other parts of the body. (Chap. 12.)

Migration

One of the evolutionary forces which may act on a population, thereby changing the frequency of genes in its gene pool. Gene flow, the migration of genes between populations, is an important factor in maintaining populational genetic variability. (Chap. 16.)

Mitochondria

A kind of cytoplasmic organelle which converts chemical substances derived from food into small waste products, a process which requires oxygen and provides the cell with energy for performing work. (Chap. 4.)

Mitosis

The form of cell division in which the chromosome number per cell is maintained at a constant level. That is, mitosis produces two daughter cells each with the same chromosome number as the original cell. Compare: *meiosis.* (Chap. 4.)

Molecule

A group of atoms held together by chemical bonds. (Chap. 7.)

Monoculture

The practice of planting one kind of crop exclusively on a broad scale. (Chap. 18.)

Monomorphism

The condition in which a population is genetically uniform for a particular trait. Contrast: *polymorphism.* (Chap. 11.)

Monozygotic

Referring to identical twins, i.e., those derived from a single fertilized egg. Contrast: *dizygotic.* (Chap. 14.)

Mosaic

A genetic mosaic is an individual who has different chromosomal or genic constitutions in different tissues of the body. (Chaps. 6, 20.)

Multiple alleles

The condition in which more than two alleles exist in a population for a particular gene. (Chap. 11.)

Muscular dystrophy

One form of this disease is inherited as an X-linked recessive trait. It is characterized by muscle degeneration, (Chaps. 10, 15.)

Mutagen

Any agent that induces mutation e.g., x-rays, certain chemicals. (Chap. 13.)

Mutation

Genetic change, either at the chromosomal or DNA level. A sudden, stable change of one allele to another (point mutation) or of one chromosomal arrangement to another (chromosomal mutation). (Chap. 3.)

Mutation rate

The rate at which genetic changes occur in a gene, usually expressed as mutations per gamete per gene per generation. (Chap. 13.)

Natural selection

The term used by Darwin to describe the process of differential survival and reproduction among individuals with different phenotypes. These selective differences produce changes in gene frequency in the population over time, i.e., bring about evolution. (Chap. 17.)

Nitrogenous bases

The genetic coding components of DNA and RNA: adenine, guanine, cytosine, thymine (DNA only), and uracil (RNA only). (Chap. 7).

Nondisjunction

The nonseparation of paired chromosomes during meiosis or mitosis. Contrast: *segregation.* (Chap. 6.)

Nucleic acids

DNA and RNA. (Chap. 7.)

Nucleotide

The phosphate-sugar-base component of nucleic acids. (Chap. 7.)

Nucleus

The cellular compartment interior to the cytoplasm, bounded by the nuclear membrane. Contains the chromosomes and other minor components. (Chap. 4.)

Oocyte

The female reproductive cell at the stage of meiosis. The post-meiotic, mature egg cell is sometimes called an ovum. (Chap. 5.)

Operator

The portion of a chromosome adjacent to a structural gene to which repressor protein and RNA polymerase can attach. (Chap. 9.)

Operon

A contiguous group of structural genes whose transcription is jointly controlled by the same regulator protein. (Chap. 9.)

Organelles

The compartmentalized subunits of the cell, such as the mitochondria. (Chap. 4.)

Overdominance

The condition in which, in a heterozygote, the phenotypic effect of the two contrasting alleles is greater than in the case of either type of homozygote. Compare: *hybrid vigor.* (App. A.)

Pedigree

A diagram of the ancestors and other relatives of an individual. (Chap. 3.)

Penetrance

Variable penetrance describes the situation in which the genotype for a trait shows an effect in some individuals and not in others. If there is no effect in an individual, the trait is said to be nonpenetrant in that case; if all relevant genotypes express the trait, this is called 100 per cent penetrance. (Chap. 15.)

Phage (bacteriophage)

Viruses that parasitize bacterial cells. (Chap. 12.)

Phenotype

The observable traits of an individual, determined to some extent by his/her genotype; more specifically, the outward expression of a particular gene or combination of genes. (Chap. 2.)

Phenylketonuria (PKU)

An inborn error of metabolism in which the normal conversion of the amino acid phenylalanine to other critical substances is blocked as the result of a deficiency of the relevant enzyme. (Chap. 10.)

Photosynthesis

The process of capturing solar energy in a chemical form useful for living systems. Occurs in chloroplasts in green plants. (Chap. 7.)

Pleiotrophy

Multiple effects of the action of a gene. (Chap. 14.)

Polygenic

Refers to a combined effect of many genes. (Chap. 14.)

Polymerase

Any of several enzymes which catalyze the polymerization of nucleotides into nucleic acid strands; e.g., RNA polymerase specifically catalyzes RNA synthesis. (Chap. 8.)

Polymorphism

The condition in which a population is genetically variable for a given trait; to be considered a polymorphism the relevant allele frequencies must be higher than can be explained by the mutation rate. Compare: *balanced polymorphism, monomorphism.* (Chaps. 13, 17.)

Polyploid

Having an excess number of chromosome sets above the diploid level of two sets. See: *triploid, tetraploid.* (Chap. 6.)

Porphyria

A metabolic condition in which chemical substances known as porphyrins are excreted in excess amounts in the urine. (Chap. 15.)

Prokaryote

Any organism in which the cell nucleus does not have a nuclear membrane; generally, bacteria and related forms. Contrast: *eukaryote.* (Chap. 9.)

Prophase

The stage in mitosis and meiosis which precedes metaphase. In prophase of meiosis I the homologous chromosomes pair, cross over (usually), and contract to minimum size. In prophase of mitosis and of meiosis II the chromosomes contract but do not, as a rule, pair or cross over. (Chap. 4.)

Proteins

Molecules composed of amino acid sequences. Proteins serve as enzymes and other facilitators of cell activity, each kind of protein serving a specific function. Proteins are gene products which are synthesized on ribosomes via mRNA instructions. (Chap. 7.)

Race

The various groups of humanity, variously defined by different people. Races often differ in allele frequencies for polymorphic genes. (Chap. 19.)

Radiation

Electromagnetic waves, e.g., x-rays, ultraviolet light, visible light. (Chap. 13.)

Random mating

Choice of a mate without regard to genotypic similarities or dissimilarities. (Chap. 16.)

Recessiveness

The condition in which, in a heterozygote, the action of one allele is masked by the phenotypic effect of the other allele. The masked allele is said to be recessive. Contrast: *dominance.* (Chap. 2.)

Reciprocal cross

Dual crosses of the type AA ♀ × aa ♂ and aa ♀ × AA ♂. Such reciprocal crosses are performed for the purpose of ascertaining if genetic factors behave similarly in both sexes. (Chap. 2.)

Regulator gene
A gene the product of which acts as a regulator of RNA transcription for other (structural) genes. (Chap. 9.)

Rem
The unit of radiation dose absorbed by human tissue. (Chap. 13.)

Replication
The process of DNA synthesis by Watson-Crick base pairing between the template strand and the components of the new replica strand. (Chap. 8.)

Repressor
A chemical substance which, by interaction with the operator, inhibits the transcription of mRNA from a gene. Repressor on the operator site can also combine with an inducer molecule, thereby relieving the inhibition of transcription. (Chap. 9.)

Retinoblastoma
A dominantly inherited eye cancer. (Chaps. 6, 15.)

Rh
A blood-group system which forms a genetic polymorphism in most human populations. The Rh system is of medical importance in the case of maternal-fetal incompatability resulting from the mother having the phenotype Rh-negative and the fetus having the phenotype Rh-positive. (Chap. 11.)

Ribosomal RNA (rRNA)
The transcript of genes specialized for rRNA production. Ribosomal RNA is of several kinds, all of which are components of ribosomes. (Chap. 8.)

Ribosomes
The small cytoplasmic entities, consisting of rRNA and proteins, to which mRNA and tRNA amino acid complexes attach to effect protein synthesis there. (Chap. 4.)

RNA (ribonucleic acid)
The cellular transcript of the genetic material (DNA). RNA occurs in a variety of forms in cells: messenger, transfer, and ribosomal. RNA is also the genetic material of some viruses. (Chap. 8.)

Schistosomiasis
This tropical disease is caused by bodily contact with waterways infested by the parasite of certain snails, leading to human infection by the parasite. (Chap. 13.)

Schizophrenia
A mental disorder, often serious, which is characterized by thought defects and other behavioral problems. (Chap. 14.)

Segregation
The process of separation of alleles of a single gene to different gametes during meiosis; discovered by Mendel. (Chap. 2.)

Self-fertilization
Fertilization of the female gametes of a plant by the male gametes of the same plant. Contrast: *cross-fertilization*. (Chap. 2.)

Sex chromosomes
The chromosomes involved in sex determination, specifically the X and Y chromosomes. (Chap. 6.)

Sibs, siblings
Brothers and sisters. (Chap. 3.)

Sickle cell anemia
The medical condition characterized by deformed red blood cells and associated symptoms: anemia, etc. An autosomal recessive trait caused by a variant protein, sickle cell hemoglobin. (Chaps. 8, 10, 15, 17, 19.)

Somatic cells
Body cells other than those with reproductive functions. (Chap. 5.)

Somatic mutation
Mutation in a somatic cell; hence, not heritable by the next generation. (Chaps. 12, 13.)

Spermatocyte
The male reproductive cell at the stage of meiosis. The post-meiotic mature reproductive cell is called a sperm. (Chap. 5.)

Species
An interbreeding group of organisms; in the traditional concept, a species is a *kind* of plant or animal. (Chap. 16.)

Structural gene
A gene from which mRNA is transcribed for translation into protein. (Chap. 9.)

Syndrome
The association of several medical abnormalities in the same individual. (Chap. 6.)

Tay-Sachs disease
An autosomal recessive condition resulting from an upset in lipid metabolism, producing infantile mental degeneration and death at an early age. (Chap. 10.)

Telophase
The stage in mitosis and meiosis which follows anaphase. In telophase the chromosomes reach the two poles of the cell and the nuclear membrane re-forms around each of these daughter nuclear masses. (Chap. 4.)

Tetraploid
Having four sets of chromosomes. (Chap. 6.)

Thalassemia
A severe disorder of red blood cells characterized by a lack, or paucity, of beta (or alpha) chains in an individual's hemoglobin molecules; an autosomal recessive trait. Normally, most hemoglobin molecules in the body, after birth, are composed of two alpha and two beta chains. (Chap. 10.)

Transcription
The transfer of genetic information coded in the base sequence of DNA to the (complementary) base sequence of RNA. (Chap. 8.)

Transfer RNA (tRNA)

The transcript of genes specialized for tRNA production. There are at least 20 kinds of tRNA molecules per cell, each kind adapted to transfer one kind of amino acid to a growing protein chain during translation of mRNA. (Chap. 8.)

Transformation by DNA

The change in genetic information in a cell by the inclusion of DNA from a different organism. Compare: *cell transformation*. (Chap. 7.)

Translation

The transfer of genetic information coded in the triplet sequence of mRNA into the amino acid sequence of protein. The process is mediated by tRNA and occurs on the ribosomes. (Chap. 8.)

Translocation

The transfer of a chromosomal fragment to a different chromosome. Transfers may be reciprocal between two chromosomes. (Chap. 6.)

Triploid

Having three sets of chromosomes. (Chap. 6.)

Trisomy

Possessing an extra chromosome in an otherwise diploid cell; thus, one kind of chromosome is represented by three homologs, instead of the usual two. (Chap. 6.)

True-breeding

Homozygosity for a given trait, thereby producing offspring uniformly like the parents upon self-fertilization or breeding among relatives. (Chap. 2.)

Turner's syndrome

The medical designation for the set of characteristics found in individuals who possess one X chromosome and no Y chromosome. (Chap. 6.)

Ultraviolet (UV)

Light of wavelengths shorter than those in the visible range but longer than the wavelengths of x-rays. (Chaps. 12, 13.)

Virus

A non-cellular life form composed of nucleic acid (DNA or RNA) and a small number of proteins. Viruses are parasites which can multiply only within cells. Many viruses cause disease. (Chap. 12.)

Watson-Crick model

The model envisions DNA as consisting of two chains wound together in a double helix. The chains are complementary to each other in base sequence and complementary bases pair together to hold the two chains in alignment. (Chap. 7.)

Wild type

The common allele in those cases where a gene also has a rare variant allele, in contrast to genes where two or more alleles are at polymorphic frequencies and none of them is uniquely common. (Chap. 17.)

Xeroderma pigmentosum

The recessive autosomal condition characterized by extreme sensitivity to sunlight and thus a susceptibility to skin cancer. A defect in repair of DNA underlies these symptoms. (Chap. 12.)

X-inactivation

The permanent condensation of one X chromosome in a genetically inactive state in XX, XXY, and other individuals who possess at least one X chromosome. Consult: *Barr body*. (Chap. 9.)

X-linkage

The location of particular genes on the X chromosome. (Chap. 5.)

XO

See *Turner's syndrome*.

XXY

See *Klinefelter's syndrome*.

XYY

The designation for men who have one X chromosome and two Y chromosomes. (Chap. 6.)

Zygote

The fertilized egg. (Chap. 2.)

Appendix C

Answers to selected study questions.

2–1. Yellow.   1/2 Yy + 1/2 YY.

2–2. Tall.   1/4.   1/4.   1:2:1.   Dwarf.

2–4. Low. Hh, low, and hh, high.

3–2. 3/4 normal, 1/4 albino.

4–3. 92.

5–3   (a). Tt Yy. (b). Tt Yy × Tt Yy. (c). 1/16.

5–4.   1/2 Ab + 1/2 ab and 1/2 aB + 1/2 ab.
1/4 Aa Bb + 1/4 aa Bb + 1/4 Aa bb +
1/4 aa bb.

5–6.   1/2 chance for having a son times 1/2 chance
for colorblindness = 1/4. Very low, in general.
Yes, the second answer depends on the genotype of the father.

6–1. Nondisjunction of X and Y in the father.

6–2. XY.

7–2. Ultimately, the sun.

7–4   (a). Gaps are shown in the sequence of bases.
(b). T is opposite G.

8–2. (e)

8–3. (c)

8–4. (d)

8–5. (c)

8–6. (b)

8–8. (d)

9–3. Regulation.

9–4. (e)

9–5. (c) or (d)

9–6. (f)

10–6. (2)

11–1. (c)

11–2. (c)

11–5. Genetic.

14–1. (d)

16–1. (b)

16–2. (e)

16–3. (e)

16–4. (e)

16–5. (d)

16–7. (b)

17–1. (d)

17–2. (e)

17–3. (b)

17–4. (b)

INDEX